Anatomy Unveiled: A Comprehensive Journey into the Human Body

Welcome to "Anatomy Unveiled: A Comprehensive Journey into the Human Body." In the pages of this book, we embark on an extraordinary expedition into the intricate and awe-inspiring realm of human anatomy. The human body is a marvel of complexity, an intricate symphony of interconnected systems and structures that work harmoniously to sustain life. This book is your guide to understanding the inner workings of the human body, from the macroscopic level down to the cellular and molecular details.

Anatomy has fascinated and captivated curious minds for centuries. From ancient explorations of the human body's structure to modern advancements in medical imaging and technology, our understanding of anatomy has grown exponentially. The study of anatomy is not only essential for medical professionals and researchers but also holds immense value for anyone curious about the intricate machinery that powers human life.

In "Anatomy Unveiled," we will delve deep into the systems that govern our body's functions, from the cardiovascular and respiratory systems that keep us alive, to the musculoskeletal system that allows us to move and interact with the world. We will explore the delicate network of nerves that transmit signals throughout our body, the intricate pathways of blood vessels that transport nutrients, and the incredible organization of cells and tissues that form the foundation of life.

Throughout this journey, we will utilize a comprehensive approach, blending text, illustrations, diagrams, and photographs to provide you with a multi-dimensional understanding of human anatomy. Whether you're a student, a healthcare professional, or simply someone curious about the human body, "Anatomy Unveiled" offers an immersive and enlightening exploration into the wonders that lie within.

So, let us embark on this educational adventure together, as we peel back the layers of skin to uncover the intricate architecture that defines us as human beings. Prepare to be amazed, inspired, and enlightened as we unveil the captivating world that lies beneath our skin.

Introduction: The Fascination of Human Anatomy
- Historical significance of anatomical studies
- The relevance of human anatomy in modern times

Chapter 1: Foundations of Anatomy
- Brief overview of anatomical terminology
- Introduction to anatomical planes, directions, and positions
- Importance of anatomical terminology in effective communication

Chapter 2: Skeletal System
- Overview of the human skeleton
- Types of bones and their functions
- Bone formation, growth, and remodeling
- Interaction between bones and muscles

Chapter 3: Muscular System

- Types of muscles (skeletal, smooth, cardiac)
- Muscle structure and contraction mechanism
- Role of muscles in movement and support
- Muscular system's contribution to metabolism and body temperature regulation

Chapter 4: Cardiovascular System

- Anatomy and function of the heart
- Blood vessels and their types (arteries, veins, capillaries)
- Blood circulation and its role in transporting nutrients and oxygen
- Heartbeat, blood pressure, and cardiovascular health

Chapter 5: Respiratory System

- Structures of the respiratory system (lungs, trachea, bronchi)
- Mechanism of respiration and gas exchange
- Role of the respiratory system in maintaining acid-base balance
- Respiratory disorders and their impact on health

Chapter 6: Nervous System

- Overview of the nervous system (central, peripheral)
- Structure and function of neurons and neurotransmitters
- Brain anatomy and major regions
- Role of the nervous system in sensation, movement, and cognition

Chapter 7: Digestive System

- Organs of the digestive system (mouth, esophagus,

stomach, intestines)
- Digestive processes and nutrient absorption
- Importance of enzymes and hormones in digestion
- Common digestive disorders and their management

Chapter 8: Endocrine System
- Major endocrine glands and their hormones
- Regulation of bodily functions through hormone secretion
- Interplay between the endocrine and nervous systems
- Endocrine disorders and their impact on health

Chapter 9: Reproductive System
- Male and female reproductive anatomy
- Reproductive processes, fertilization, and pregnancy
- Hormonal regulation of the reproductive system
- Reproductive health and family planning

Chapter 10: Integumentary System
- Structure and functions of the skin
- Importance of the skin as a protective barrier
- Role of sweat glands, hair, and nails in maintaining homeostasis
- Skin disorders and care

Chapter 11: Sensory Systems
- Visual system and the anatomy of the eye
- Auditory system and the anatomy of the ear
- Gustatory and olfactory senses
- Role of sensory systems in perception and communication

Chapter 12: Putting It All Together: Interconnected Systems
- Integration of different body systems for overall functioning
- Case studies highlighting the interactions between systems
- The importance of holistic understanding in medical practice

Conclusion: Unveiling the Wonders Within
- Reflection on the journey through human anatomy
- The ongoing relevance of anatomical knowledge
- Encouragement for further exploration and appreciation of the human body

Appendices
- Glossary of anatomical terms
- Resources for additional reading and learning

Historical significance of anatomical studies

Anatomical studies have played a crucial and lasting role in the development of scientific knowledge, medical advancements, and our understanding of the human body. The historical significance of anatomical studies can be summarized as follows:

1. **Ancient Civilizations:** Even in ancient civilizations like Egypt, Greece, and Rome, anatomical studies were conducted to gain insights into the human body's structure and function. Early anatomical observations laid the foundation for medical practices and beliefs.

2. **Renaissance and the Age of Enlightenment:** The Renaissance marked a resurgence of interest in human anatomy. Pioneers like Leonardo da Vinci conducted detailed anatomical dissections, creating accurate illustrations of the human body's internal structures. These studies marked the transition from relying on ancient authorities to empirical observation and scientific inquiry.

3. **Andreas Vesalius and the Birth of Modern Anatomy:** Andreas Vesalius, a Flemish anatomist in the 16th century, revolutionized anatomical studies with his groundbreaking work "De Humani Corporis Fabrica." This meticulously detailed book presented accurate illustrations and descriptions of human anatomy, challenging many earlier misconceptions.

4. **Advancements in Medical Education:** Accurate anatomical knowledge became essential for medical education and training. Anatomical dissections provided students with practical insights into the

body's structures and functions, enabling more effective diagnosis and treatment.

5. **Impact on Medical Practice:** Anatomical studies have directly impacted medical practice by enabling the development of surgical techniques, interventions, and medical innovations. Knowledge of anatomy is fundamental for performing surgeries, administering medications, and understanding disease mechanisms.

6. **Contributions to Physiology and Pathology:** Anatomical studies have contributed to our understanding of how different organs and systems function. By studying anatomical variations and anomalies, scientists gained insights into both normal and abnormal physiological processes.

7. **Discovery of Diseases:** Anatomical studies have led to the discovery and understanding of various diseases and conditions. Identifying structural changes in organs and tissues helped elucidate the causes and effects of illnesses, leading to advancements in medical diagnostics and treatment.

8. **Evolution of Medical Illustration:** The detailed illustrations created during anatomical studies have become valuable historical records and educational resources. These illustrations have influenced medical art, education, and scientific communication.

9. **Ethical and Social Considerations:** Throughout history, anatomical studies raised ethical and social concerns. The use of cadavers for dissections, often obtained from deceased individuals, sparked debates about respect for the dead and ethical treatment of human remains.

10. **Continued Relevance:** Anatomical studies continue to be a cornerstone of medical education, research, and innovation. Technological advancements, such as medical imaging and virtual dissections, have expanded the ways in which anatomy is studied and applied.

In summary, anatomical studies have shaped the course of medical science, education, and practice throughout history. From ancient civilizations to modern medicine, these studies have provided a deep understanding of the human body's intricate structures and functions, driving advancements that improve healthcare, save lives, and enhance our overall knowledge of the human condition.

The relevance of human anatomy in modern times

Human anatomy remains highly relevant in modern times due to its crucial role in various fields, including medicine, biology, healthcare, forensics, education, and technology. Here's why human anatomy continues to be relevant:

1. **Medical Diagnosis and Treatment:**
   - An understanding of human anatomy is essential for accurate medical diagnosis and treatment. Physicians need to know the body's structures to identify and address health issues effectively.

2. **Surgical Procedures:**
   - Surgeons rely on detailed anatomical knowledge to perform surgeries safely and successfully. Accurate surgical planning and execution require a deep understanding of the body's structures and their relationships.

3. **Medical Imaging and Diagnostics:**
   - Modern medical imaging technologies, such as MRI, CT scans, and ultrasound, generate images based on anatomical structures. Interpreting these images requires expertise in anatomy.

4. **Biomedical Research:**
   - Anatomical knowledge underpins biomedical research, enabling scientists to study disease mechanisms, develop therapies, and

understand how the body functions at a molecular level.

5. **Pharmaceutical Development:**
   - Understanding the body's anatomy is crucial for designing drugs that target specific anatomical structures, such as receptors and enzymes.

6. **Medical Education:**
   - Medical students learn anatomy as the foundation of their medical education. A solid grasp of anatomy is essential for becoming competent healthcare professionals.

7. **Healthcare Professions:**
   - Nurses, physical therapists, radiologists, and other healthcare professionals require anatomical knowledge to provide quality patient care.

8. **Forensic Science:**
   - Forensic experts use anatomical knowledge to identify remains, determine causes of death, and provide critical evidence in legal investigations.

9. **Educational Curricula:**
   - Anatomy is a core subject in educational curricula for aspiring healthcare professionals, helping them build a strong foundation for their careers.

10. **Medical Illustration and Animation:**
    - Medical illustrators and animators create visual representations of anatomical structures for educational materials, textbooks, and patient education.

11. **Virtual Reality and Simulation:**
    - Virtual reality and simulation technologies enable learners to explore the human body in a three-dimensional, interactive manner,

enhancing education and training.

12. Prosthetics and Implants:

- The design and fitting of prosthetics and implants require a deep understanding of anatomy to ensure optimal function and integration with the body.

13. Bioengineering and Biotechnology:

- Innovations in biotechnology and bioengineering, such as tissue engineering and organ transplantation, rely on anatomical knowledge.

14. Art and Aesthetics:

- Artists, sculptors, and animators use anatomical references to create realistic depictions of the human form in art and entertainment.

15. Public Health and Wellness:

- Public health initiatives, disease prevention, and wellness promotion benefit from understanding the anatomical basis of health and disease.

In essence, human anatomy serves as the foundation upon which advancements in various fields are built. Its relevance continues to extend into modern times, shaping the way we understand the human body, innovate in healthcare, and improve overall well-being.

Brief overview of anatomical terminology

Anatomical terminology is a standardized language used to describe the structures and relationships within the human body. It provides a precise and universally understood way to communicate about anatomy. Here's a brief overview of anatomical terminology:

1. **Anatomical Position:**
   - The standard reference position for describing anatomical structures.
   - The body is upright, facing forward, with arms at the sides and palms forward, and feet parallel.
2. **Planes of the Body:**
   - Sagittal Plane: Divides the body into left and right halves.
   - Frontal (Coronal) Plane: Divides the body into front (anterior) and back (posterior) portions.
   - Transverse (Horizontal) Plane: Divides the body into upper (superior) and lower (inferior) parts.
3. **Directional Terms:**
   - Superior: Above or toward the head.
   - Inferior: Below or toward the feet.
   - Anterior (ventral): Toward the front.
   - Posterior (dorsal): Toward the back.
   - Medial: Toward the midline.
   - Lateral: Away from the midline.
   - Proximal: Closer to the trunk or point of origin.
   - Distal: Farther from the trunk or point of

origin.

4. **Body Regions:**
   - Abdominal: Area between the chest and hips.
   - Thoracic: Chest region.
   - Pelvic: Lower abdominal area.
   - Cranial: Skull region.
   - Caudal: Tailbone or coccyx area.

5. **Body Landmarks:**
   - Epidermis: Outermost layer of skin.
   - Dorsal and Palmar: Back and palm of the hand, respectively.
   - Plantar: Sole of the foot.
   - Antecubital and Popliteal: Front of the elbow and back of the knee, respectively.

6. **Body Cavities:**
   - Cranial Cavity: Contains the brain.
   - Thoracic Cavity: Contains the heart, lungs, and major blood vessels.
   - Abdominal Cavity: Contains organs such as the liver, stomach, and intestines.
   - Pelvic Cavity: Contains reproductive organs, bladder, and rectum.

7. **Body Systems:**
   - Each major body system (e.g., skeletal, muscular, cardiovascular) has its own set of anatomical terms to describe its components and functions.

8. **Terms for Movement:**
   - Flexion: Bending a joint to decrease the angle between bones.
   - Extension: Straightening a joint to increase the angle between bones.
   - Abduction: Moving a limb away from the midline.
   - Adduction: Moving a limb toward the midline.
   - Rotation: Turning a bone on its own axis.

- Pronation: Turning the palm downward.
- Supination: Turning the palm upward.

9. **Terms for Relationships and Composition:**
  - Superficial: Near the surface.
  - Deep: Farther from the surface.
  - Proximal and Distal: Used to describe the relationship between two structures along a limb.
  - External and Internal: Referring to positions outside or inside a structure.

Anatomical terminology provides a common language for healthcare professionals, researchers, educators, and anyone studying the human body. It enables precise communication, accurate documentation, and a clear understanding of anatomical structures and relationships.

Introduction to anatomical planes, directions, and positions

Anatomical planes, directions, and positions are fundamental concepts used to describe the human body's structure and relationships. They provide a standardized language that healthcare professionals, anatomists, and researchers use to communicate accurately and effectively about anatomical features. Let's explore these concepts in more detail:

Anatomical Planes: Anatomical planes are imaginary flat surfaces used to visualize and describe the body's divisions. There are three primary anatomical planes:

1. **Sagittal Plane:** This vertical plane divides the body into left and right halves. If the plane passes exactly through the midline, it's called the midsagittal plane; if it's offset from the midline, it's referred to as a parasagittal plane.
2. **Frontal (Coronal) Plane:** This vertical plane divides the body into front (anterior) and back (posterior) portions. It is perpendicular to the sagittal plane.
3. **Transverse (Horizontal) Plane:** This horizontal plane divides the body into upper (superior) and lower (inferior) parts. It is perpendicular to both the sagittal and frontal planes.

Anatomical Directions: Anatomical directions are terms used to describe the relative position of structures within the body:

1. **Superior:** Refers to a structure being above another structure or closer to the head (cranial).

2. **Inferior:** Refers to a structure being below another structure or closer to the feet (caudal).
3. **Anterior (Ventral):** Refers to a structure being toward the front of the body.
4. **Posterior (Dorsal):** Refers to a structure being toward the back of the body.
5. **Medial:** Describes a structure that is closer to the midline of the body.
6. **Lateral:** Describes a structure that is farther away from the midline.
7. **Proximal:** Describes a structure that is closer to the trunk or point of origin.
8. **Distal:** Describes a structure that is farther away from the trunk or point of origin.

Anatomical Positions: Anatomical positions are standardized reference points for describing the body's orientation:

1. **Anatomical Position:** The body is upright, facing forward, with arms at the sides and palms forward, and feet parallel. This position provides a consistent starting point for anatomical descriptions.
2. **Prone Position:** The body is lying face down.
3. **Supine Position:** The body is lying face up.

Understanding these concepts is crucial for accurately describing and interpreting anatomical structures and relationships. They serve as the foundation for effective communication in medical and scientific contexts, facilitating clear and precise discussions about the human body's complex organization.

Importance of anatomical terminology in effective communication

Anatomical terminology plays a pivotal role in effective communication within the medical, scientific, and healthcare communities. Its importance lies in its ability to provide a standardized and universally understood language for describing the complex structures, relationships, and orientations of the human body. Here are several reasons highlighting the significance of anatomical terminology in effective communication:

1. **Precision and Clarity:** Anatomical terminology offers precise and specific terms that leave no room for ambiguity or misinterpretation. This clarity is essential when discussing intricate anatomical features, positions, and relationships.

2. **Global Understanding:** Anatomical terminology follows standardized conventions that are recognized and used internationally. This ensures that medical professionals, researchers, and educators from different countries can communicate seamlessly using the same terminology.

3. **Accurate Documentation:** In medical practice, accurate documentation of patient conditions, procedures, and findings is crucial. Anatomical terminology ensures that medical records are consistent and detailed, promoting accurate diagnosis and treatment.

4. **Educational Consistency:** Anatomical terminology provides a consistent framework for teaching and

learning about the human body. Educators can convey information with precision, and students can grasp complex concepts more effectively.

5. **Clinical Communication:** In healthcare settings, clear communication among healthcare professionals is essential for patient safety and quality care. Anatomical terms enable effective communication between doctors, nurses, surgeons, and other specialists.

6. **Medical Imaging Interpretation:** Medical imaging reports, such as X-rays, CT scans, and MRIs, often include anatomical descriptors. Radiologists and clinicians rely on these terms to accurately interpret and communicate imaging findings.

7. **Research Collaboration:** Anatomical terminology facilitates collaboration among researchers in various fields, including anatomy, physiology, pathology, and biomechanics. A common language enables effective exchange of ideas and findings.

8. **Diagnostic Accuracy:** Anatomical terms aid in the precise description of patient symptoms and physical findings. This accuracy contributes to accurate diagnoses and appropriate treatment plans.

9. **Surgical Planning:** Surgeons use anatomical terminology to plan and communicate about surgical procedures. Clear communication ensures that surgical teams are aligned and fully understand the planned intervention.

10. **Legal and Ethical Contexts:** Anatomical terminology is important in legal contexts, such as medical malpractice cases or autopsy reports. Accurate terminology helps establish facts and avoid misunderstandings.

11. **Professional Development:** Healthcare professionals, medical students, and researchers need a solid grasp of anatomical terminology to excel in their careers and contribute effectively to their respective fields.

12. **Public Communication:** Even when communicating

with the general public, anatomical terminology can help provide accurate health information, promote understanding of medical conditions, and enhance health literacy.

In summary, anatomical terminology serves as a bridge of understanding among professionals and stakeholders in the medical and scientific realms. Its standardized language ensures clarity, consistency, and accuracy in communication, contributing to improved patient care, medical research, and education.

Overview of the human skeleton

The human skeleton is the framework of bones that provides structure, support, protection, and movement for the body. Comprising 206 bones at adulthood, the skeleton can be categorized into two main parts: the axial skeleton and the appendicular skeleton.

Axial Skeleton: The axial skeleton forms the central axis of the body and includes the following components:

1. **Skull:** The skull consists of the cranium (protecting the brain) and the facial bones. It also houses structures like the eye sockets, nasal passages, and oral cavity.
2. **Vertebral Column (Spine):** The vertebral column consists of individual vertebrae stacked upon each other. It surrounds and protects the spinal cord, providing flexibility and support.
3. **Rib Cage:** The rib cage includes the ribs, sternum (breastbone), and thoracic vertebrae. It protects the heart and lungs and is involved in respiration.

Appendicular Skeleton: The appendicular skeleton comprises the bones of the limbs and their associated girdles. It allows for movement and supports the body's actions:

1. **Upper Limbs:**
   - Shoulder Girdle: Includes the scapula (shoulder blade) and clavicle (collarbone), which attach the upper limbs to the axial skeleton.
   - Arm: Composed of the humerus bone, the upper arm supports movement and

attachment of muscles.

- Forearm: The radius and ulna bones make up the forearm, allowing rotation and movement of the wrist and hand.
- Hand: The hand includes the carpals (wrist bones), metacarpals (hand bones), and phalanges (finger bones).

2. **Lower Limbs:**
   - Pelvic Girdle: Consists of the hip bones (ilium, ischium, and pubis) and supports the lower limbs while also protecting the pelvic organs.
   - Thigh: The femur is the longest bone in the body and forms the thigh.
   - Leg: The tibia (shinbone) and fibula form the leg, supporting body weight and allowing for ankle movements.
   - Foot: The foot includes tarsal bones (ankle bones), metatarsals (foot bones), and phalanges (toe bones).

In addition to providing structural support and movement, the human skeleton has various other functions:

- **Hematopoiesis:** Some bones, such as the ribs, vertebrae, and long bones, house bone marrow where blood cells are produced.
- **Mineral Storage:** Bones store essential minerals, including calcium and phosphorus, which are vital for maintaining bone health and overall bodily functions.
- **Protection:** Bones protect vital organs like the brain, heart, and lungs from injury.
- **Muscle Attachment:** Bones serve as anchor points for muscles, enabling movement and body actions.

Overall, the human skeleton is a complex and dynamic system that is integral to the body's structure, function, and overall well-being.

Types of bones and their functions

Bones in the human body can be classified into several types based on their shape and function. Each type of bone serves specific purposes and contributes to the overall structure and function of the skeleton. The main types of bones and their functions are as follows:

1. **Long Bones:**
   - Examples: Femur, humerus, radius, ulna, tibia, fibula.
   - Function: Long bones are responsible for providing support, facilitating movement, and acting as levers for muscular action. They also contribute to bone marrow production, particularly in the central cavities of long bones.

2. **Short Bones:**
   - Examples: Carpals (wrist bones), tarsals (ankle bones).
   - Function: Short bones provide stability, support, and limited movement. They also play a role in shock absorption and weight distribution.

3. **Flat Bones:**
   - Examples: Skull bones, scapula, sternum, ribs.
   - Function: Flat bones protect internal organs, provide broad surfaces for muscle attachment, and contribute to the formation of blood cells in their red bone marrow.

4. **Irregular Bones:**

- Examples: Vertebrae, facial bones, pelvis.
- Function: Irregular bones have complex shapes that help protect internal organs, provide attachment points for muscles, and support bodily functions.

5. **Sesamoid Bones:**
   - Examples: Patella (kneecap).
   - Function: Sesamoid bones develop within tendons and help to reduce friction and increase mechanical advantage during joint movement.

6. **Accessory Bones:**
   - Examples: Wormian bones in the skull.
   - Function: Accessory bones are smaller, extra bones that sometimes develop in certain individuals. Their function can vary, but they may contribute to joint stability or have no functional significance.

7. **Pneumatic Bones:**
   - Examples: Bones of the skull containing air-filled cavities, such as the frontal, ethmoid, and sphenoid bones.
   - Function: Pneumatic bones contribute to the structure of the skull and may help reduce the weight of the head.

The functions of bones go beyond their classification, as bones collectively support and protect the body, facilitate movement, provide attachment points for muscles and ligaments, and participate in physiological processes such as blood cell production and mineral storage. The various types of bones work together to create a dynamic and resilient skeletal system that enables humans to carry out a wide range of activities.

Bone formation, growth, and remodeling

Bone formation, growth, and remodeling are dynamic processes that occur throughout a person's life. These processes are crucial for maintaining bone health, adapting to changing physiological demands, and repairing bone injuries. Here's an overview of each process:

Bone Formation (Ossification): Bone formation, also known as ossification, involves the development of bone tissue from precursor cells. There are two main types of bone ossification:

1. **Intramembranous Ossification:**
   - Occurs primarily in flat bones like the skull and clavicle.
   - Mesenchymal cells differentiate into osteoblasts, which secrete bone matrix.
   - The bone matrix mineralizes, forming trabeculae or flat bones.
2. **Endochondral Ossification:**
   - Occurs in most bones, including long bones.
   - Cartilage model forms first and is gradually replaced by bone tissue.
   - Cartilage cells (chondrocytes) hypertrophy and die, allowing osteoblasts to deposit bone matrix in their place.

Bone Growth: Bone growth occurs primarily during childhood and adolescence as a result of cell division and the addition of bone tissue to existing bones.

1. **Longitudinal Growth:**

- Occurs at the epiphyseal plates (growth plates) located at the ends of long bones.
- Chondrocytes in the growth plates divide, increasing the length of the bone.
- Osteoblasts replace the cartilage with bone tissue.

2. **Appositional Growth:**
   - Bone growth in width or thickness.
   - Osteoblasts on the bone's surface deposit new bone tissue, while osteoclasts remove old tissue.

Bone Remodeling: Bone remodeling is an ongoing process that involves the removal of old bone tissue and the deposition of new bone tissue. It helps maintain bone health, adapt bones to mechanical stresses, and repair microdamage. The process is regulated by the coordinated activity of bone-resorbing cells (osteoclasts) and bone-forming cells (osteoblasts).

1. **Resorption Phase:**
   - Osteoclasts break down and remove old bone tissue.
   - Minerals and degraded matrix components are released into the bloodstream.

2. **Formation Phase:**
   - Osteoblasts deposit new bone tissue, replacing the resorbed bone.
   - New bone tissue is laid down in layers and organized along lines of mechanical stress.

Bone remodeling ensures that bones maintain their strength, adapt to changing mechanical loads, and repair minor damage. It is influenced by various factors, including hormonal regulation, physical activity, calcium levels, and overall health.

Throughout life, the processes of bone formation, growth, and remodeling work together to maintain bone health, structural integrity, and the body's ability to support movement and protect vital organs.

Interaction between bones and muscles

The interaction between bones and muscles is a fundamental aspect of the body's biomechanics and allows for movement, stability, and coordination. This interaction occurs at specialized junctions known as "muscle-bone attachments" or "muscle insertions." The interaction between bones and muscles can be understood through the following key points:

1. Muscle Contraction and Movement:

- Muscles contract and generate force, which is transmitted to the bones through tendons, tough connective tissue structures that attach muscles to bones.
- Muscle contraction causes movement at joints, resulting in actions like flexion, extension, rotation, and more.

2. Tendon Attachments:

- Tendons connect muscles to bones and play a critical role in transmitting the force generated by muscle contraction to the bones.
- Tendons are strong and durable to withstand the mechanical stress generated during muscle contraction.

3. Muscle Origin and Insertion:

- The origin of a muscle is the point where it attaches to a relatively stationary bone.
- The insertion of a muscle is the point where it attaches to a bone that moves as a result of muscle contraction.
- Contraction of a muscle causes movement by pulling the

insertion toward the origin.

4. Muscle-Antagonist Pairs:

- Muscles usually work in pairs or groups to produce coordinated movement.
- The agonist (prime mover) muscle contracts to cause a specific movement, while the antagonist muscle relaxes.
- The antagonist muscle opposes the movement generated by the agonist and helps control the movement's speed and precision.

5. Stabilization and Joint Support:

- Muscles play a critical role in stabilizing joints and maintaining posture.
- Certain muscles, called stabilizers, contract to prevent excessive movement or dislocation of joints during dynamic activities.

6. Lever System and Mechanical Advantage:

- Bones act as levers, and joints act as fulcrums, in the musculoskeletal system.
- Different lever systems (first, second, and third class) provide mechanical advantages that influence the force and speed of movement.

7. Adaptation and Training:

- Regular physical activity and strength training can lead to adaptations in both muscles and bones.
- Muscles can become stronger and more efficient in generating force, while bones can increase in density and strength in response to mechanical stress.

The interaction between bones and muscles is a coordinated and complex process that allows the body to perform a wide range of movements and activities. This relationship is essential for everyday functions, athletic performance, and overall health.

Types of muscles (skeletal, smooth, cardiac)

There are three main types of muscles in the human body: skeletal muscle, smooth muscle, and cardiac muscle. Each type of muscle has unique characteristics, functions, and locations within the body:

1. Skeletal Muscle:

- Also known as voluntary or striated muscle.
- Location: Attached to bones by tendons and responsible for movement of the body.
- Appearance: Striated or striped appearance under a microscope due to the arrangement of protein filaments.
- Control: Under conscious control of the nervous system, allowing for voluntary movement.
- Function: Skeletal muscles enable movement of bones and joints, maintain posture, and generate heat.
- Example: Muscles that control limb movement, facial expressions, and posture.

2. Smooth Muscle:

- Also known as involuntary or non-striated muscle.
- Location: Found in the walls of internal organs, blood vessels, and other structures.
- Appearance: Lack striations and have a smooth appearance under a microscope.
- Control: Involuntary control by the autonomic nervous system.
- Function: Smooth muscles regulate movement in

various organs, such as the digestive system, blood vessels, and respiratory tract.

- Example: Muscles in the walls of the stomach, intestines, and blood vessels.

3. Cardiac Muscle:

- Unique type of muscle found only in the heart.
- Location: Forms the heart's walls, known as the myocardium.
- Appearance: Striated like skeletal muscle but with branching and intercalated discs that allow for synchronized contractions.
- Control: Involuntary control by the autonomic nervous system, but the heart can also generate its own electrical impulses.
- Function: Cardiac muscles contract rhythmically to pump blood throughout the body.
- Characteristics: Highly resistant to fatigue due to continuous, rhythmic contractions.

These three types of muscles have distinct structures and functions that are suited to their specific roles within the body. Skeletal muscles provide movement and support, smooth muscles regulate organ function, and cardiac muscles ensure the continuous pumping of blood to sustain life.

Muscle structure and contraction mechanism

Muscle structure and contraction mechanism are essential aspects of understanding how muscles generate force and movement in the body. The structure of muscles is intricately linked to their ability to contract, allowing for various physiological processes. Here's an overview of muscle structure and the mechanism of muscle contraction:

Muscle Structure:

1. **Muscle Fibers (Cells):**
   - Muscles are composed of elongated cells called muscle fibers.
   - Muscle fibers are multinucleated and contain specialized cellular structures called myofibrils.

2. **Myofibrils:**
   - Myofibrils are cylindrical structures within muscle fibers that contain the contractile proteins responsible for muscle contraction.
   - Each myofibril is divided into repeating units called sarcomeres.

3. **Sarcomeres:**
   - Sarcomeres are the functional units of muscle contraction.
   - They consist of alternating bands of dark and light regions, giving rise to the striated appearance of skeletal and cardiac muscles.

4. **Myofilaments:**
   - Myofilaments are the proteins that make up the sarcomeres.

- Thick filaments are made of the protein myosin, while thin filaments are composed of the protein actin.

Mechanism of Muscle Contraction:

1. **Sliding Filament Theory:**
   - Muscle contraction occurs through the sliding filament theory, which describes the interaction between thick and thin filaments within sarcomeres.

2. **Cross-Bridge Formation:**
   - During muscle contraction, myosin heads (extensions of myosin molecules) form cross-bridges with actin molecules in the thin filaments.

3. **Calcium Ion Release:**
   - Muscle contraction is triggered by the release of calcium ions from the sarcoplasmic reticulum (a specialized organelle within muscle cells).

4. **Troponin and Tropomyosin:**
   - Calcium ions bind to troponin, causing a conformational change that moves tropomyosin away from the binding sites on actin.

5. **Power Stroke:**
   - When calcium ions bind to troponin, myosin heads bind to actin molecules, forming cross-bridges.
   - Myosin heads undergo a power stroke, causing them to pivot and pull the thin filaments toward the center of the sarcomere.

6. **ATP and Muscle Relaxation:**
   - ATP (adenosine triphosphate) is required for muscle contraction.
   - When ATP binds to myosin heads, the cross-

bridges detach, allowing the muscle to relax.

7. **Role of Calcium Ions:**
   - Calcium ions play a key role in regulating muscle contraction by binding to troponin and initiating the contraction process.

8. **Role of Nervous System:**
   - The nervous system stimulates muscle contraction by sending electrical signals (action potentials) to motor neurons, which release neurotransmitters that activate muscle fibers.

Overall, the mechanism of muscle contraction involves the interaction between myofilaments, the release of calcium ions, and the energy provided by ATP. This process allows muscles to generate force and movement, contributing to various activities and functions within the body.

Role of muscles in movement and support

Muscles play a vital role in both movement and support within the human body. They work in coordination with bones, joints, and other structures to enable a wide range of movements and maintain the body's stability and posture. Here's how muscles contribute to movement and support:

Role of Muscles in Movement:

1. **Generating Force:** Muscles generate force through contraction, allowing them to pull on bones and create movement at joints. This force generation is essential for all types of body movements.
2. **Muscle Contractions:** Muscles can contract in different ways to produce various types of movements:
   - Concentric Contraction: Muscles shorten as they generate force, resulting in movement.
   - Eccentric Contraction: Muscles lengthen under tension, controlling movement or resisting external forces.
   - Isometric Contraction: Muscles contract without changing length, providing stability or maintaining posture.
3. **Agonist and Antagonist Muscles:** Muscles often work in pairs or groups to produce coordinated movements. The agonist (prime mover) muscle contracts to create a specific movement, while the antagonist muscle relaxes to allow the movement to occur smoothly.
4. **Muscle Synergists:** Synergist muscles assist the agonist muscle in creating movements, helping to stabilize joints and optimize movement patterns.

5. **Complex Movements:** Muscles and muscle groups work together to produce complex movements involving multiple joints and muscles. Examples include walking, running, jumping, and lifting objects.

Role of Muscles in Support and Stability:

1. **Postural Support:** Muscles are essential for maintaining an upright posture. They counteract the force of gravity to prevent the body from collapsing under its own weight.
2. **Joint Stabilization:** Muscles provide stability to joints by holding bones in proper alignment and preventing excessive movement. This is particularly important in weight-bearing joints.
3. **Muscle Tone:** Even at rest, muscles maintain a certain level of tension known as muscle tone. Muscle tone provides a baseline level of support to maintain posture and prevent joints from becoming too loose or too stiff.
4. **Joint Protection:** Muscles surrounding joints act as protective buffers, absorbing shocks and impacts to prevent joint damage.
5. **Dynamic Stability:** Muscles provide dynamic stability during movements by adjusting their level of contraction to control joint movement and prevent sudden shifts.
6. **Core Stability:** Muscles in the core region, including the abdominal and back muscles, play a crucial role in stabilizing the spine and pelvis during various activities.

In summary, muscles are essential for movement and support in the human body. They enable a wide range of motions, from simple tasks to complex athletic movements, while also maintaining posture, protecting joints, and providing stability. The intricate interplay between muscles, bones, and joints ensures that the body can function effectively and efficiently in various environments and situations.

Muscular system's contribution to metabolism and body temperature regulation

The muscular system plays significant roles in metabolism and body temperature regulation due to the energy demands and heat production associated with muscle contraction. These functions are closely interconnected and essential for maintaining homeostasis within the body. Here's how the muscular system contributes to metabolism and body temperature regulation:

Contribution to Metabolism:

1. **Energy Expenditure:** Muscle contractions require energy, and the process of converting chemical energy (from nutrients) into mechanical work is called metabolism. Muscles are metabolically active tissues that consume a substantial amount of energy during contraction and relaxation.

2. **Resting Metabolic Rate (RMR):** Muscles have a higher resting metabolic rate compared to other tissues. Even at rest, they require energy for maintenance, repair, and the maintenance of muscle tone.

3. **Caloric Expenditure:** Physical activity, including exercise and everyday movements, increases caloric expenditure due to the energy required for muscle contractions. Muscles are major contributors to total energy expenditure.

4. **Fat Utilization:** Muscle contractions rely on energy sources such as glucose and fatty acids. Regular exercise can enhance the body's ability to utilize stored fats

for energy, contributing to weight management and metabolic health.

5. **Muscle Growth and Protein Synthesis:** Building and repairing muscle tissue require protein synthesis, which involves energy expenditure. Adequate protein intake and regular exercise support muscle growth and maintenance.

Contribution to Body Temperature Regulation:

1. **Heat Production:** Muscle contractions generate heat as a byproduct of metabolic processes. During exercise or physical activity, muscle contractions increase heat production, contributing to the body's overall heat output.

2. **Shivering:** When the body's core temperature drops, muscle contractions can occur involuntarily in a rapid and rhythmic manner, leading to shivering. Shivering generates heat to help restore normal body temperature.

3. **Basal Metabolic Rate (BMR):** Muscles have a higher metabolic rate than other tissues even at rest. This constant energy expenditure generates heat and contributes to maintaining body temperature.

4. **Exercise-Induced Thermogenesis:** Physical activity and exercise increase body temperature due to increased metabolic activity in muscles. This phenomenon is known as exercise-induced thermogenesis.

5. **Cooling Mechanisms:** During exercise, increased blood flow to muscles facilitates heat dissipation to the skin's surface. Sweating and dilation of blood vessels in the skin help dissipate excess heat, preventing overheating.

The muscular system's contributions to metabolism and body temperature regulation are essential for the body's overall function and well-being. By generating energy, producing heat, and supporting thermoregulation, muscles play critical roles in maintaining the body's internal balance and responding to various physiological demands.

Anatomy and function of the heart

The heart is a vital organ that pumps blood throughout the body, delivering oxygen and nutrients to cells and removing waste products. It is located within the chest cavity and is responsible for maintaining the circulation of blood to sustain life. Here's an overview of the anatomy and function of the heart:

Anatomy of the Heart:

1. **Chambers:** The heart has four chambers: two atria (upper chambers) and two ventricles (lower chambers). The right atrium receives deoxygenated blood from the body, while the left atrium receives oxygenated blood from the lungs. The right ventricle pumps deoxygenated blood to the lungs for oxygenation, and the left ventricle pumps oxygenated blood to the rest of the body.
2. **Valves:** Four valves ensure one-way blood flow through the heart:
   - Tricuspid Valve: Separates the right atrium and right ventricle.
   - Pulmonary Valve: Separates the right ventricle from the pulmonary artery.
   - Mitral Valve (Bicuspid Valve): Separates the left atrium and left ventricle.
   - Aortic Valve: Separates the left ventricle from the aorta.
3. **Septum:** The heart is divided into left and right halves by a muscular partition called the septum.
4. **Coronary Arteries:** These arteries supply the heart muscle (myocardium) with oxygenated blood. Blockages in these arteries can lead to coronary artery disease.

Function of the Heart:

1. **Blood Circulation:** The heart serves as a pump that propels blood through a closed circulatory system. Deoxygenated blood returns to the heart from the body, is oxygenated in the lungs, and then pumped out to the body's tissues.

2. **Pulmonary Circulation:** Deoxygenated blood from the body enters the right atrium, passes through the tricuspid valve, and then enters the right ventricle. The right ventricle pumps blood through the pulmonary valve into the pulmonary artery, which carries the blood to the lungs for oxygenation.

3. **Systemic Circulation:** Oxygenated blood returns from the lungs to the left atrium, passes through the mitral valve, and enters the left ventricle. The left ventricle pumps blood through the aortic valve into the aorta, which distributes the oxygenated blood to all the body's tissues.

4. **Cardiac Cycle:** The cardiac cycle consists of systole (contraction) and diastole (relaxation) phases. These cycles ensure blood is pumped efficiently and effectively.

5. **Heart Rate and Rhythm:** The heart rate is the number of times the heart contracts per minute. The heart's rhythm is controlled by electrical signals from the sinoatrial (SA) node, also known as the heart's natural pacemaker.

6. **Blood Pressure Regulation:** The heart plays a role in regulating blood pressure by adjusting its pumping force based on the body's needs and maintaining adequate circulation.

7. **Heart Sounds:** Heart sounds (lub-dub) are produced by the closing of heart valves during the cardiac cycle and are indicative of the heart's functioning.

In summary, the heart's anatomy and function are intricately linked to its role in pumping blood to supply oxygen and nutrients to the body's tissues. Through the coordination of chambers, valves, and circulatory pathways, the heart ensures efficient circulation, oxygenation, and overall cardiovascular health.

Blood vessels and their types
(arteries, veins, capillaries)

Blood vessels are tubular structures that transport blood throughout the body, facilitating the circulation of oxygen, nutrients, hormones, and waste products. There are three main types of blood vessels: arteries, veins, and capillaries. Each type has unique characteristics and functions in the circulatory system:

1. Arteries: Arteries are thick-walled blood vessels that carry oxygenated blood away from the heart to various parts of the body. Arteries have several layers, including an inner endothelium, a middle layer of smooth muscle, and an outer connective tissue layer. Key features of arteries include:

- **Elasticity:** Arteries have elastic fibers in their walls that allow them to stretch and recoil, helping to maintain blood pressure during cardiac cycles.
- **Muscularity:** The smooth muscle layer can contract and regulate blood flow by adjusting the diameter of the arteries, a process known as vasoconstriction or vasodilation.
- **Pressure Reservoir:** Arteries serve as pressure reservoirs that propel blood forward and maintain blood flow during diastole (relaxation phase of the heart).

2. Veins: Veins are blood vessels that carry deoxygenated blood back to the heart from various parts of the body. Veins have thinner walls than arteries and contain valves to prevent backflow of blood. Characteristics of veins include:

- **Valves:** Valves within veins prevent blood from flowing backward (against gravity) and help maintain unidirectional blood flow toward the heart.
- **Large Capacitance:** Veins have a larger capacity to hold blood, acting as blood reservoirs that can accommodate changes in blood volume.
- **Low Pressure:** Blood pressure in veins is lower than in arteries due to the lower resistance to blood flow.

3. Capillaries: Capillaries are the smallest and thinnest blood vessels, connecting arterioles (small arteries) to venules (small veins). They facilitate the exchange of nutrients, gases, and waste products between blood and tissues. Characteristics of capillaries include:

- **Microscopic Size:** Capillaries have a very small diameter, allowing them to penetrate tissues and deliver nutrients and oxygen to cells.
- **Single-Cell Layer:** Capillary walls are composed of a single layer of endothelial cells, allowing for efficient exchange of substances between blood and tissues.
- **Slow Blood Flow:** Blood flow through capillaries is slow, facilitating the exchange of substances and maximizing diffusion.

Collectively, arteries, veins, and capillaries work together to ensure the proper functioning of the circulatory system. Arteries carry oxygenated blood to tissues, veins return deoxygenated blood to the heart, and capillaries facilitate the exchange of substances between blood and tissues. This complex network of blood vessels plays a critical role in maintaining the body's overall health and homeostasis.

Blood circulation and its role in transporting nutrients and oxygen

Blood circulation is the continuous movement of blood through the circulatory system, which includes the heart, blood vessels, and lungs. Circulation is essential for transporting nutrients, oxygen, hormones, and waste products to and from the body's cells. The circulation of blood occurs in two main pathways: pulmonary circulation and systemic circulation.

1. Pulmonary Circulation: Pulmonary circulation involves the circulation of blood between the heart and the lungs. Its main function is to oxygenate the blood and remove carbon dioxide. The pathway of pulmonary circulation is as follows:

- Deoxygenated blood from the body enters the right atrium of the heart through the superior and inferior vena cava.
- The right atrium contracts, sending blood through the tricuspid valve into the right ventricle.
- The right ventricle contracts, pumping blood through the pulmonary valve into the pulmonary artery.
- The pulmonary artery carries deoxygenated blood to the lungs, where oxygen is added and carbon dioxide is removed through the process of gas exchange in the alveoli.
- Oxygenated blood returns to the heart through the pulmonary veins and enters the left atrium.

2. Systemic Circulation: Systemic circulation involves the circulation of oxygenated blood to the body's tissues and the return of deoxygenated blood to the heart. The pathway of systemic circulation is as follows:

- Oxygenated blood from the left atrium enters the left ventricle.
- The left ventricle contracts, pumping blood through the aortic valve into the aorta, the largest artery.
- The aorta branches into smaller arteries that carry oxygenated blood to various parts of the body.
- In the capillaries, oxygen and nutrients are exchanged for carbon dioxide and waste products in the tissues.
- Deoxygenated blood is collected in venules and small veins, which merge into larger veins.
- The deoxygenated blood returns to the heart through the superior and inferior vena cava, entering the right atrium to start the cycle again.

Role in Transporting Nutrients and Oxygen:

- Oxygen: Oxygen-rich blood from the lungs is pumped by the left ventricle into the aorta and distributed to all tissues through systemic circulation. Oxygen diffuses from capillaries into cells, where it is used in cellular respiration to produce energy.
- Nutrients: Nutrients obtained from digested food are absorbed into the bloodstream through the digestive system. The bloodstream carries these nutrients to cells throughout the body for energy production, growth, and repair.
- Gas Exchange: In the lungs, oxygen from inhaled air diffuses into the bloodstream, while carbon dioxide diffuses out of the bloodstream and is exhaled.

Overall, blood circulation plays a critical role in ensuring that oxygen and nutrients are delivered to cells and that waste products, including carbon dioxide, are removed. This process is essential for maintaining the body's cellular functions, energy production, and overall health.

Heartbeat, blood pressure, and cardiovascular health

Heartbeat, blood pressure, and cardiovascular health are closely interconnected aspects of the circulatory system that play a crucial role in maintaining overall well-being. Here's an overview of each of these elements and their significance for cardiovascular health:

1. Heartbeat (Cardiac Cycle): The heartbeat, or cardiac cycle, refers to the rhythmic contraction and relaxation of the heart's chambers, which allows blood to be pumped throughout the body. The cardiac cycle consists of systole (contraction) and diastole (relaxation) phases. The heart's natural pacemaker, the sinoatrial (SA) node, generates electrical signals that initiate each heartbeat. The heartbeat is regulated by the autonomic nervous system and hormones. A regular heartbeat is essential for maintaining effective blood circulation and ensuring the body's oxygen and nutrient supply.

2. Blood Pressure: Blood pressure is the force exerted by circulating blood against the walls of blood vessels. It is essential for driving blood through the circulatory system. Blood pressure is measured in millimeters of mercury (mmHg) and is recorded as two values: systolic pressure (the higher value) and diastolic pressure (the lower value). Blood pressure can fluctuate based on various factors, including physical activity, stress, and hormonal changes. Elevated blood pressure, known as hypertension, can strain the heart and blood vessels, increasing the risk of cardiovascular diseases.

3. Cardiovascular Health: Cardiovascular health refers to the overall well-being of the heart, blood vessels, and circulatory system. Maintaining good cardiovascular health is crucial for optimal functioning and longevity. Factors that contribute to cardiovascular health include:

- **Healthy Diet:** Consuming a balanced diet rich in fruits, vegetables, whole grains, lean proteins, and healthy fats can support heart health by providing essential nutrients and antioxidants.
- **Regular Physical Activity:** Regular exercise helps improve cardiovascular fitness, maintain a healthy weight, and enhance circulation.
- **Blood Pressure Management:** Monitoring and managing blood pressure within a healthy range helps reduce the risk of cardiovascular diseases.
- **Cholesterol Levels:** Maintaining healthy cholesterol levels, including low levels of LDL (bad) cholesterol and higher levels of HDL (good) cholesterol, is essential for heart health.
- **Smoking Cessation:** Quitting smoking reduces the risk of cardiovascular diseases and improves overall health.
- **Stress Management:** Managing stress through relaxation techniques, exercise, and mindfulness can positively impact cardiovascular health.
- **Regular Checkups:** Regular medical checkups and screenings help detect and address cardiovascular risk factors and diseases early.

Cardiovascular diseases, such as heart disease and stroke, are among the leading causes of death globally. Promoting cardiovascular health through healthy lifestyle choices and regular medical care is essential for reducing the risk of these conditions and maintaining a strong, functioning circulatory system.

Structures of the respiratory system (lungs, trachea, bronchi)

The respiratory system is responsible for the exchange of oxygen and carbon dioxide between the body and the environment. It includes various structures that work together to facilitate breathing and the exchange of gases. Here are the main structures of the respiratory system:

1. Lungs: The lungs are the primary organs of the respiratory system and are responsible for the exchange of oxygen and carbon dioxide. The right lung has three lobes, while the left lung has two lobes to accommodate the heart. The lungs are enclosed by a double-layered membrane called the pleura, which helps reduce friction during breathing.

2. Trachea: The trachea, also known as the windpipe, is a tube that connects the larynx (voice box) to the bronchi. It is composed of cartilage rings that provide structural support and prevent the trachea from collapsing during inhalation. The trachea allows air to pass from the upper respiratory tract to the lower respiratory tract.

3. Bronchi: The trachea branches into two bronchi, one leading to each lung. These bronchi are further divided into smaller bronchioles, forming the bronchial tree. Bronchioles continue to divide into even smaller structures called alveolar ducts, which eventually lead to tiny air sacs known as alveoli.

4. Alveoli: Alveoli are the small, thin-walled air sacs located at the end of the bronchioles within the lungs. These sacs are surrounded by a network of capillaries. Gas exchange occurs in the alveoli, where oxygen from inhaled air diffuses into the bloodstream, and carbon dioxide, a waste product of metabolism, diffuses from the blood into the alveoli to be exhaled.

5. Diaphragm: The diaphragm is a dome-shaped muscle that separates the chest cavity (thoracic cavity) from the abdominal cavity. It plays a critical role in breathing by contracting and relaxing. During inhalation, the diaphragm contracts and moves downward, increasing the volume of the thoracic cavity and allowing the lungs to expand. During exhalation, the diaphragm relaxes, causing it to move upward and reducing the lung volume.

6. Larynx: The larynx, or voice box, is located at the top of the trachea. It contains the vocal cords, which vibrate when air passes through them, producing sound. The larynx also helps prevent foreign objects from entering the trachea and lungs by closing off during swallowing.

7. Pharynx: The pharynx, or throat, is a muscular tube that serves as a passage for both air and food. It connects the nasal cavity and mouth to the trachea and esophagus. The pharynx has three sections: nasopharynx, oropharynx, and laryngopharynx.

These structures work together to ensure the proper functioning of the respiratory system, enabling the exchange of gases and supporting the body's oxygen supply and waste gas elimination.

Mechanism of respiration and gas exchange

Respiration is the process of exchanging gases (oxygen and carbon dioxide) between the body and the environment. It involves both external respiration, which occurs in the lungs, and internal respiration, which takes place at the cellular level. The mechanism of respiration and gas exchange can be understood through the following steps:

1. Ventilation: Ventilation is the process of moving air in and out of the lungs. It includes two phases: inhalation (inspiration) and exhalation (expiration).

- **Inhalation:** During inhalation, the diaphragm contracts and moves downward, while the intercostal muscles between the ribs contract, causing the ribcage to expand. This increases the volume of the thoracic cavity, creating a negative pressure. Air rushes into the lungs to equalize the pressure, bringing oxygen-rich air into the alveoli.
- **Exhalation:** Exhalation is a passive process in which the diaphragm and intercostal muscles relax. The elastic recoil of the lung tissues and the chest wall decreases the thoracic cavity's volume. As a result, air is pushed out of the lungs, expelling carbon dioxide.

2. Gas Exchange in the Alveoli: Gas exchange occurs in the alveoli, where oxygen from inhaled air diffuses across the alveolar walls into the capillaries surrounding the alveoli. Carbon dioxide, a waste product of metabolism, diffuses from the capillaries into the alveoli to be exhaled.

3. Transportation of Gases: Oxygen binds to hemoglobin in red blood cells, forming oxyhemoglobin, which is transported to body tissues. Carbon dioxide binds to hemoglobin and is transported as carbaminohemoglobin. Most of the oxygen is carried in this form, with a small amount dissolved directly in the plasma.

4. Internal Respiration: In the body's tissues, internal respiration takes place, where oxygen is delivered to cells, and carbon dioxide is produced as a waste product of cellular metabolism. Oxygen diffuses from the capillaries into cells, where it is used in cellular respiration to produce energy. Carbon dioxide produced in cells diffuses into the bloodstream and is transported back to the lungs for exhalation.

5. Gas Exchange at the Tissues: At the tissues, the opposite gas exchange process occurs compared to the alveoli. Oxygen diffuses from the capillaries into cells, and carbon dioxide diffuses from the cells into the capillaries.

The mechanism of respiration and gas exchange ensures a constant supply of oxygen to body tissues and the removal of carbon dioxide, allowing cells to function properly and maintain homeostasis. This process is essential for providing energy for cellular activities and eliminating waste products.

Role of the respiratory system in maintaining acid-base balance

The respiratory system plays a crucial role in maintaining the body's acid-base balance, also known as the pH balance. This balance is essential for proper physiological function and maintaining overall health. The respiratory system regulates acid-base balance through the control of carbon dioxide (CO_2) levels, which influences the concentration of bicarbonate ions (HCO_3^-) in the blood.

1. Carbon Dioxide and Bicarbonate Ion Relationship: The primary mechanism through which the respiratory system regulates acid-base balance involves the relationship between carbon dioxide and bicarbonate ions in the blood. Carbon dioxide is produced as a waste product of cellular metabolism and combines with water to form carbonic acid (H_2CO_3). Carbonic acid can dissociate into hydrogen ions (H^+) and bicarbonate ions (HCO_3^-):

$$CO_2 + H_2O \rightleftharpoons H_2CO_3 \rightleftharpoons H^+ + HCO_3^-$$

2. Respiratory Regulation of Acid-Base Balance: The respiratory system helps regulate acid-base balance in the following ways:

- **Control of CO2 Levels:** The rate and depth of breathing, controlled by the brainstem's respiratory centers, can be adjusted to increase or decrease the elimination of carbon dioxide from the body. Faster and deeper breathing (hyperventilation) eliminates more CO_2, leading to a decrease in carbonic acid and hydrogen ions, which raises blood pH (alkalosis). Slower and shallower breathing (hypoventilation) retains more CO_2, leading to an increase in carbonic acid and hydrogen ions, which lowers blood pH (acidosis).

- **Compensation:** If there is an imbalance in the blood's acid-base status (such as metabolic acidosis or alkalosis), the respiratory system can help compensate by adjusting ventilation. For example, if blood is too acidic (low pH), the respiratory system can increase ventilation to reduce carbon dioxide and raise the pH.

3. Interaction with Kidneys: The kidneys also play a significant role in regulating acid-base balance through the excretion of hydrogen ions and the reabsorption of bicarbonate ions. The respiratory system and the kidneys work together to maintain the body's pH within a narrow range to prevent acidosis (excess acid) or alkalosis (excess base).

4. Acid-Base Imbalances: Respiratory acidosis occurs when there is an accumulation of carbon dioxide in the blood due to hypoventilation. This leads to an increase in hydrogen ions and a decrease in blood pH. Respiratory alkalosis occurs when there is excessive elimination of carbon dioxide due to hyperventilation, resulting in a decrease in hydrogen ions and an increase in blood pH.

In summary, the respiratory system's regulation of carbon dioxide levels and acid-base balance is essential for maintaining the body's pH within a narrow range. By adjusting ventilation, the respiratory system helps prevent the buildup of carbonic acid and hydrogen ions, which can lead to acidosis, or their depletion, which can lead to alkalosis. This collaboration between the respiratory system and other regulatory mechanisms ensures the body's optimal functioning and homeostasis.

Respiratory disorders and their impact on health

Respiratory disorders are medical conditions that affect the normal functioning of the respiratory system, which includes the lungs, airways, and other structures involved in breathing. These disorders can have a significant impact on health, causing symptoms that range from mild discomfort to severe respiratory distress. Some common respiratory disorders and their impact on health include:

1. Asthma: Asthma is a chronic inflammatory disorder of the airways, characterized by wheezing, coughing, shortness of breath, and chest tightness. During asthma attacks, the airways become narrowed due to inflammation and bronchoconstriction, making it difficult to breathe. Asthma attacks can be triggered by allergens, exercise, cold air, and respiratory infections. Proper management through medications and avoiding triggers is essential to prevent severe attacks and maintain lung function.

2. Chronic Obstructive Pulmonary Disease (COPD): COPD is a group of progressive lung diseases, including chronic bronchitis and emphysema, that cause airflow limitation and difficulty breathing. COPD is often caused by long-term exposure to irritants such as cigarette smoke, air pollution, and occupational hazards. It leads to persistent cough, excessive mucus production, shortness of breath, and reduced exercise tolerance. COPD is a leading cause of disability and mortality worldwide.

3. Pneumonia: Pneumonia is an infection of the lungs that causes inflammation in the air sacs. It can result from bacterial, viral, fungal, or other types of infections. Symptoms include fever, chills, cough, chest pain, and difficulty breathing. Pneumonia can be especially severe in older adults and those with weakened immune systems. Treatment involves antibiotics for bacterial pneumonia and supportive care for viral pneumonia.

4. Pulmonary Embolism: A pulmonary embolism occurs when a blood clot, usually originating in the legs, travels to the lungs and blocks a pulmonary artery. This condition can cause sudden shortness of breath, chest pain, and rapid heart rate. It is a medical emergency that requires prompt treatment to prevent further complications.

5. Lung Cancer: Lung cancer is the uncontrolled growth of abnormal cells in the lung tissues. It is often associated with a history of smoking but can also occur in non-smokers. Symptoms include persistent cough, chest pain, weight loss, and difficulty breathing. Early detection and treatment are crucial for improving outcomes.

6. Respiratory Infections (Flu, Common Cold, COVID-19): Respiratory infections, including influenza (flu), the common cold, and COVID-19, can cause symptoms such as cough, congestion, sore throat, fever, and difficulty breathing. While most cases are mild, severe infections can lead to respiratory distress and pneumonia, especially in vulnerable populations.

7. Sleep Apnea: Sleep apnea is a sleep disorder characterized by interruptions in breathing during sleep. It can lead to loud snoring, choking or gasping for breath, and daytime fatigue. Sleep apnea is associated with an increased risk of cardiovascular diseases and other health issues.

The impact of respiratory disorders on health can vary widely depending on the severity of the condition, the individual's overall health, and the availability of appropriate treatment. Respiratory disorders can limit physical activities, decrease quality of life, and increase the risk of complications. Early diagnosis, proper management, and adopting a healthy lifestyle can help mitigate the impact of these disorders and improve overall respiratory health.

Overview of the nervous system (central, peripheral)

The nervous system is a complex network of cells, tissues, and organs that coordinate and regulate the body's functions, respond to stimuli, and transmit information throughout the body. It plays a vital role in controlling various bodily processes, including movement, sensation, thought, and behavior. The nervous system is divided into two main parts: the central nervous system (CNS) and the peripheral nervous system (PNS).

1. Central Nervous System (CNS): The central nervous system consists of the brain and spinal cord. It serves as the main control center for processing and integrating sensory information, initiating motor responses, and facilitating higher cognitive functions. Key features of the CNS include:

- **Brain:** The brain is the command center of the nervous system. It is responsible for processing sensory input, initiating motor responses, regulating emotions, and supporting cognitive functions such as thinking, memory, and decision-making. The brain is divided into different regions with specific functions, including the cerebral cortex, cerebellum, and brainstem.
- **Spinal Cord:** The spinal cord is a long, thin bundle of nerve tissue that extends from the base of the brain down the back. It serves as a conduit for transmitting nerve signals between the brain and the rest of the body. The spinal cord also plays a role in reflex actions, which are rapid responses to sensory stimuli that do not

involve conscious thought.

2. Peripheral Nervous System (PNS): The peripheral nervous system consists of nerves and ganglia (collections of nerve cell bodies) outside the CNS. It connects the CNS to the rest of the body, allowing for communication between the central nervous system and various organs, muscles, and tissues. The PNS can be further divided into two main components: the somatic nervous system and the autonomic nervous system.

- **Somatic Nervous System:** The somatic nervous system is responsible for voluntary movements and conscious perception of sensory stimuli. It consists of sensory neurons that transmit information from sensory receptors (such as the skin and muscles) to the CNS and motor neurons that transmit commands from the CNS to muscles, enabling voluntary movements.

- **Autonomic Nervous System (ANS):** The autonomic nervous system regulates involuntary functions that are essential for maintaining homeostasis, such as heart rate, digestion, and respiratory rate. The ANS is further divided into the sympathetic and parasympathetic divisions, which have opposing effects on various bodily functions. The sympathetic division prepares the body for "fight or flight" responses, while the parasympathetic division promotes "rest and digest" functions.

The nervous system functions through the transmission of electrical signals called nerve impulses or action potentials. Neurons, specialized cells of the nervous system, transmit these signals by receiving, processing, and transmitting information. Glial cells, or neuroglia, support and protect neurons by providing structural and metabolic support.

Overall, the nervous system is a sophisticated network that coordinates bodily functions, enables communication between different parts of the body, and allows us to interact with our environment and respond to stimuli.

Structure and function of neurons and neurotransmitters

Neurons are specialized cells that are the building blocks of the nervous system. They are responsible for transmitting electrical signals, called nerve impulses or action potentials, to facilitate communication within the nervous system. Neurons have unique structures that enable them to perform their functions effectively.

Structure of Neurons: Neurons have three main parts:

1. **Cell Body (Soma):** The cell body contains the nucleus and most of the cell's organelles. It integrates incoming signals from dendrites and, based on this information, generates nerve impulses.
2. **Dendrites:** Dendrites are branched extensions that receive signals from other neurons or sensory receptors. They play a role in transmitting signals toward the cell body.
3. **Axon:** The axon is a long, thin extension of the neuron that carries nerve impulses away from the cell body. It is covered by a myelin sheath, a fatty insulating layer that speeds up signal transmission and helps maintain the electrical signal's strength.
4. **Axon Terminals (Synaptic Terminals):** At the end of the axon, there are small swellings called axon terminals or synaptic terminals. These terminals contain vesicles filled with neurotransmitters, which are chemical messengers that transmit signals to other neurons or target cells.

Function of Neurons: Neurons are responsible for transmitting signals within the nervous system. The process involves the following steps:

1. **Signal Reception:** Dendrites receive incoming signals from other neurons or sensory receptors. These signals can be either excitatory (encouraging the neuron to fire) or inhibitory (preventing the neuron from firing).
2. **Integration:** The cell body integrates the incoming signals from dendrites. If the overall signal reaches a certain threshold, an action potential is generated.
3. **Action Potential Generation:** An action potential is a rapid electrical impulse that travels along the axon. It is initiated by a change in the neuron's membrane potential, resulting in depolarization. This change in membrane potential is achieved through the opening and closing of ion channels.
4. **Conduction:** The action potential travels along the axon toward the axon terminals. In myelinated axons, the signal "jumps" from one node of Ranvier (unmyelinated portion) to the next, a process known as saltatory conduction.
5. **Neurotransmitter Release:** When the action potential reaches the axon terminals, it triggers the release of neurotransmitters from vesicles into the synaptic cleft (the tiny gap between neurons). These neurotransmitters bind to receptors on the membrane of the target neuron or target cell.
6. **Signal Transmission:** The binding of neurotransmitters to receptors on the target neuron's dendrites or cell body leads to excitatory or inhibitory effects, influencing whether an action potential is generated in the target neuron.

Neurotransmitters are essential for communication between neurons and for transmitting signals across synapses (the junctions between neurons). They can have excitatory or inhibitory effects, depending on the receptors they bind to. Some common neurotransmitters include dopamine, serotonin, acetylcholine, and gamma-aminobutyric acid (GABA).

In summary, neurons are specialized cells with distinct structures that allow them to transmit electrical signals and communicate with other cells. Neurotransmitters play a vital role in transmitting signals across synapses, enabling the nervous system to carry out its functions, including sensation, movement, cognition, and behavior.

Brain anatomy and major regions

The brain is a complex and highly organized organ that serves as the control center of the nervous system. It is responsible for processing information, controlling bodily functions, and facilitating various cognitive and emotional processes. The brain can be divided into several major regions, each with distinct functions. Here is an overview of the brain's anatomy and its major regions:

1. Cerebrum: The cerebrum is the largest and most highly developed part of the brain. It is divided into two hemispheres, the left and the right, which are connected by a bundle of nerve fibers called the corpus callosum. The cerebrum is responsible for higher cognitive functions, including conscious thought, perception, memory, language, and voluntary movement. It is further divided into lobes:

- **Frontal Lobe:** Located at the front of the brain, the frontal lobe is involved in decision-making, planning, motor control, and personality.
- **Parietal Lobe:** Positioned behind the frontal lobe, the parietal lobe processes sensory information, spatial awareness, and perception of touch, temperature, and pain.
- **Temporal Lobe:** Found on the sides of the brain, the temporal lobe plays a role in auditory processing, language comprehension, and memory.
- **Occipital Lobe:** Situated at the back of the brain, the occipital lobe is responsible for visual processing and interpretation.

2. Cerebellum: The cerebellum is located at the back of the brain, just below the cerebrum. It is involved in coordinating voluntary movements, balance, posture, and fine motor skills. The cerebellum receives sensory input from the body and integrates it to produce smooth and coordinated movements.

3. Brainstem: The brainstem is the lower portion of the brain that connects the cerebrum and cerebellum to the spinal cord. It controls basic functions necessary for survival, such as breathing, heart rate, digestion, and blood pressure. The brainstem consists of several structures, including the medulla oblongata, pons, and midbrain.

4. Diencephalon: The diencephalon is located between the cerebral hemispheres and the brainstem. It includes several important structures, such as the thalamus, hypothalamus, and pineal gland.

- **Thalamus:** The thalamus acts as a relay center for sensory information entering the brain and directs it to the appropriate areas of the cerebral cortex.
- **Hypothalamus:** The hypothalamus plays a critical role in regulating homeostasis, controlling the autonomic nervous system, body temperature, hunger, thirst, sleep, and endocrine functions.
- **Pineal Gland:** The pineal gland produces melatonin, a hormone that helps regulate the sleep-wake cycle.

The brain's intricate structure and organization allow it to perform a wide range of functions essential for maintaining bodily functions and supporting cognition, emotion, and behavior. The interaction between these different regions enables complex processes and behaviors.

Role of the nervous system in sensation, movement, and cognition

The nervous system plays a crucial role in various aspects of human functioning, including sensation, movement, and cognition. These functions involve complex interactions between different parts of the nervous system, particularly the brain and peripheral nervous system. Here's an overview of the role of the nervous system in sensation, movement, and cognition:

1. Sensation: Sensation refers to the process of detecting and interpreting sensory information from the environment and the body. The nervous system plays a key role in transmitting sensory signals from sensory receptors to the brain, where they are processed and interpreted. The major senses involved in sensation include:

- **Vision:** Visual sensory information is received by the eyes' photoreceptor cells, transmitted through the optic nerves, and processed in the visual cortex of the brain.
- **Hearing:** Auditory sensory information is detected by the ears' hair cells, transmitted through the auditory nerves, and processed in the auditory cortex.
- **Taste:** Gustatory sensory information is detected by taste buds on the tongue, transmitted through the facial and glossopharyngeal nerves, and processed in the gustatory cortex.
- **Smell:** Olfactory sensory information is detected by olfactory receptor cells in the nose, transmitted through the olfactory nerve, and processed in the olfactory

cortex.

- **Touch:** Tactile sensory information, including pressure, temperature, and pain, is detected by sensory receptors in the skin, transmitted through various sensory nerves, and processed in the somatosensory cortex.

2. Movement: The nervous system is responsible for coordinating voluntary and involuntary movements of the body. This involves a complex interaction between the brain, spinal cord, and peripheral nerves. Motor neurons transmit signals from the brain to muscles, causing them to contract and generate movement. The major components of movement control include:

- **Voluntary Movement:** The motor cortex in the brain initiates and controls voluntary movements. The cerebellum helps coordinate movements and ensure their accuracy and smoothness.
- **Reflexes:** Reflexes are involuntary responses to sensory stimuli that involve the spinal cord and bypass the brain. They help protect the body from harm and maintain balance.

3. Cognition: Cognition refers to the mental processes involved in acquiring knowledge, processing information, and generating thoughts, emotions, and behaviors. The nervous system, particularly the cerebral cortex, plays a central role in cognitive functions, including:

- **Memory:** The hippocampus and other brain regions are involved in forming, storing, and retrieving memories.
- **Learning:** The process of acquiring new knowledge and skills is facilitated by the integration of sensory input and cognitive processing.
- **Reasoning and Problem Solving:** The prefrontal cortex is involved in executive functions such as decision-making, planning, and problem-solving.
- **Emotion:** The limbic system, including the amygdala

and hypothalamus, is associated with emotional processing and regulation.

- **Language:** Various brain regions, such as Broca's area and Wernicke's area, are responsible for language production and comprehension.

In summary, the nervous system's intricate structure and functions enable the body to perceive the environment through sensation, generate movement, and engage in complex cognitive processes such as memory, learning, emotion, and language. The interactions between different parts of the nervous system contribute to the integrated functioning of the human body and mind.

Organs of the digestive system (mouth, esophagus, stomach, intestines)

The digestive system is responsible for breaking down food and absorbing nutrients that the body needs for energy, growth, and maintenance. It consists of a series of organs that work together to facilitate digestion and absorption. Here are the main organs of the digestive system:

1. Mouth: The mouth is where the process of digestion begins. It contains the following structures:

- **Teeth:** Teeth mechanically break down food into smaller pieces through chewing (mastication).
- **Salivary Glands:** Salivary glands secrete saliva, which contains enzymes (such as amylase) that initiate the digestion of carbohydrates. Saliva also helps moisten food and facilitate swallowing.
- **Tongue:** The tongue helps manipulate food during chewing and plays a role in forming food into a bolus (soft mass) for swallowing.

2. Esophagus: The esophagus is a muscular tube that connects the mouth to the stomach. It transports food from the mouth to the stomach through peristalsis, a series of coordinated muscle contractions.

3. Stomach: The stomach is a muscular, J-shaped organ that continues the digestion process. It has several important functions:

- **Mechanical Mixing:** The stomach's muscles mix food

with gastric juices, creating a semi-liquid mixture called chyme.

- **Acidic Environment:** Gastric juices, including hydrochloric acid, help break down proteins and kill bacteria present in food.
- **Enzyme Secretion:** The stomach secretes enzymes like pepsin to further break down proteins.

4. Small Intestine: The small intestine is the primary site of nutrient absorption. It is divided into three parts: the duodenum, jejunum, and ileum. The small intestine has specialized structures to increase its surface area for absorption:

- **Villi:** Villi are finger-like projections that line the walls of the small intestine and contain blood vessels and lymphatic vessels. They increase the surface area for nutrient absorption.
- **Microvilli:** Microvilli are even smaller projections on the surface of the villi, further enhancing nutrient absorption.

5. Large Intestine (Colon): The large intestine is responsible for absorbing water, electrolytes, and some vitamins. It also plays a role in forming and storing feces before elimination. The large intestine includes the cecum, colon, rectum, and anus.

- **Cecum:** The cecum is a pouch-like structure at the beginning of the large intestine and is the site of the appendix, a small vestigial organ with immune functions.
- **Colon:** The colon is divided into several segments, including the ascending colon, transverse colon, descending colon, and sigmoid colon. It is involved in water absorption and the formation of feces.
- **Rectum:** The rectum stores feces until they are ready to be eliminated from the body.
- **Anus:** The anus is the opening through which feces are

expelled from the body.

These organs work together to digest food, extract nutrients, and eliminate waste products, allowing the body to obtain the energy and nutrients it needs to function properly.

Digestive processes and nutrient absorption

The digestive system is responsible for breaking down food into its individual components and absorbing nutrients that the body needs for energy, growth, and various physiological processes. The digestive processes involve mechanical and chemical actions that occur in different parts of the digestive tract. Here's an overview of the digestive processes and nutrient absorption:

1. Ingestion: Ingestion is the process of taking food into the mouth and swallowing it. Chewing (mastication) breaks down food into smaller pieces, increasing its surface area for digestion.

2. Propulsion: Propulsion refers to the movement of food through the digestive tract. Peristalsis, coordinated muscular contractions, propels food from the esophagus to the stomach and through the entire length of the intestines.

3. Mechanical Digestion: Mechanical digestion involves physically breaking down food into smaller pieces. This process enhances the efficiency of chemical digestion and exposes a larger surface area for enzyme action.

4. Chemical Digestion: Chemical digestion involves the breakdown of food molecules into simpler substances through the action of enzymes. Key chemical digestion processes include:

- **Carbohydrate Digestion:** Begins in the mouth with salivary amylase and continues in the small intestine with pancreatic amylase. Carbohydrates are broken down into glucose and other simple sugars.
- **Protein Digestion:** Begins in the stomach with the enzyme pepsin and continues in the small intestine

with pancreatic enzymes (trypsin, chymotrypsin, carboxypeptidase). Proteins are broken down into amino acids.

- **Fat Digestion:** Takes place primarily in the small intestine. Bile produced by the liver emulsifies fats into smaller droplets, and pancreatic lipase breaks down fats into fatty acids and glycerol.

5. Absorption: Absorption is the process by which nutrients are taken into the bloodstream or lymphatic system for distribution to cells throughout the body. Absorption primarily occurs in the small intestine, where specialized structures increase the surface area for nutrient uptake:

- **Villi:** Finger-like projections on the lining of the small intestine that contain blood vessels and lymphatic vessels. Nutrients are absorbed through the villi into the bloodstream or lymph.
- **Microvilli:** Tiny projections on the surface of villi that further increase the surface area for absorption.

6. Nutrient Transport: After absorption, nutrients are transported by the bloodstream to cells throughout the body. Glucose and amino acids are transported through the blood, while fatty acids and fat-soluble vitamins are transported through the lymphatic system.

7. Elimination: The final step of digestion is the elimination of indigestible and unabsorbed materials as feces. This process occurs in the large intestine and involves water absorption, which solidifies the feces.

Overall, the digestive processes and nutrient absorption involve a combination of mechanical and chemical actions that occur in different parts of the digestive system. This allows the body to break down food into its basic components and absorb essential nutrients for various physiological functions.

Importance of enzymes and hormones in digestion

Enzymes and hormones play critical roles in the digestion process by facilitating the breakdown of food into smaller, absorbable molecules and regulating various digestive functions. These biochemical substances are produced by different organs and glands within the body and contribute to the efficiency of the digestive process. Here's an overview of the importance of enzymes and hormones in digestion:

Enzymes in Digestion: Enzymes are protein molecules that act as biological catalysts, accelerating chemical reactions without being consumed in the process. In digestion, enzymes are essential for breaking down complex food molecules into simpler forms that can be absorbed by the body. Different enzymes are involved in breaking down carbohydrates, proteins, and fats:

- **Amylase:** Amylase enzymes, found in saliva and pancreatic juices, break down complex carbohydrates (starches) into simple sugars (e.g., glucose, maltose).
- **Pepsin:** Pepsin, produced by the stomach, is responsible for the initial breakdown of proteins into smaller peptides.
- **Trypsin, Chymotrypsin, Carboxypeptidase:** These enzymes, produced by the pancreas and released into the small intestine, further break down proteins into amino acids.
- **Lipase:** Lipase enzymes, produced by the pancreas and released into the small intestine, break down fats

(triglycerides) into fatty acids and glycerol.

- **Pancreatic Amylase:** Pancreatic amylase continues the breakdown of carbohydrates in the small intestine.
- **Maltase, Sucrase, Lactase:** These enzymes, located on the surface of intestinal villi, further break down disaccharides (maltose, sucrose, lactose) into monosaccharides.

Hormones in Digestion: Hormones are chemical messengers produced by various glands and organs. They regulate digestion by signaling the release of digestive enzymes, controlling the movement of food, and maintaining overall digestive processes. Some important digestive hormones include:

- **Gastrin:** Produced by the stomach, gastrin stimulates the secretion of gastric juices, including hydrochloric acid and pepsin, which aid in protein digestion.
- **Cholecystokinin (CCK):** Produced by the small intestine in response to the presence of fats and proteins, CCK stimulates the release of bile from the gallbladder and enzymes from the pancreas.
- **Secretin:** Also produced by the small intestine, secretin stimulates the pancreas to release bicarbonate-rich pancreatic juices that help neutralize stomach acid and create an optimal pH environment for enzyme activity in the small intestine.
- **Gastric Inhibitory Peptide (GIP):** Produced by the small intestine, GIP inhibits gastric acid secretion and stimulates insulin release from the pancreas in response to the ingestion of nutrients.
- **Enterogastrone:** Produced by the small intestine, enterogastrone inhibits stomach emptying and gastric acid secretion, slowing down digestion when needed for optimal nutrient absorption.

Overall, enzymes and hormones work together to ensure the effective breakdown of food, the release of necessary nutrients, and the proper coordination of digestive processes. Their roles are critical for extracting energy and nutrients from the diet and maintaining overall digestive health.

Common digestive disorders and their management

Digestive disorders are conditions that affect the normal functioning of the digestive system, leading to various symptoms and discomfort. These disorders can range from mild to severe and may require different management approaches. Here are some common digestive disorders and their management:

1. Gastroesophageal Reflux Disease (GERD): GERD occurs when stomach acid flows back into the esophagus, causing symptoms like heartburn, regurgitation, and chest pain. Management includes:

- Lifestyle Modifications: Avoiding trigger foods, losing weight, raising the head of the bed, and eating smaller meals.
- Medications: Antacids, H2 blockers, proton pump inhibitors (PPIs), and prokinetic agents to reduce stomach acid and promote better digestion.
- Surgical Interventions: In severe cases, surgical procedures to strengthen the lower esophageal sphincter may be considered.

2. Irritable Bowel Syndrome (IBS): IBS is a functional disorder characterized by abdominal pain, bloating, diarrhea, and/or constipation. Management includes:

- Dietary Changes: Identifying trigger foods and following a low-FODMAP diet or other personalized dietary plans.

- Stress Management: Relaxation techniques, cognitive-behavioral therapy, and regular exercise to manage stress.
- Medications: Antispasmodics, laxatives, antidiarrheal medications, and medications targeting specific symptoms.

3. Inflammatory Bowel Disease (IBD): IBD includes conditions like Crohn's disease and ulcerative colitis, which cause chronic inflammation of the digestive tract. Management involves:

- Medications: Anti-inflammatory drugs, immunosuppressants, biologic therapies, and corticosteroids to reduce inflammation and manage symptoms.
- Lifestyle Changes: Dietary modifications, smoking cessation, and stress reduction techniques.
- Surgery: Surgical removal of affected portions of the digestive tract in severe cases.

4. Celiac Disease: Celiac disease is an autoimmune disorder triggered by gluten consumption, leading to damage in the small intestine. Management involves:

- Gluten-Free Diet: Completely avoiding foods containing gluten, including wheat, barley, and rye.
- Nutritional Support: Ensuring adequate intake of nutrients like iron, calcium, and B vitamins through gluten-free sources or supplements.

5. Gallstones: Gallstones are solid particles that form in the gallbladder and can cause pain, nausea, and vomiting. Management options include:

- Observation: Asymptomatic gallstones may not require treatment but should be monitored.
- Surgery: Surgical removal of the gallbladder (cholecystectomy) may be recommended for

symptomatic gallstones.

6. Lactose Intolerance: Lactose intolerance is the inability to digest lactose, a sugar found in dairy products. Management includes:

- Dietary Changes: Avoiding lactose-containing foods and beverages or using lactase supplements.
- Lactase Supplements: Taking lactase supplements before consuming dairy products to help digest lactose.

It's important to note that individual management plans may vary based on the severity of the disorder and the patient's specific needs. If you suspect you have a digestive disorder or experience persistent symptoms, it's recommended to seek medical advice for proper diagnosis and personalized management.

Major endocrine glands and their hormones

The endocrine system consists of various glands that secrete hormones directly into the bloodstream. These hormones play critical roles in regulating a wide range of physiological processes, including metabolism, growth, development, reproduction, and stress response. Here are some of the major endocrine glands and the hormones they produce:

1. Hypothalamus: The hypothalamus is a region of the brain that controls the release of hormones from the pituitary gland. It produces hormones called releasing and inhibiting hormones, which regulate the pituitary gland's hormone secretion.

2. Pituitary Gland (Hypophysis): The pituitary gland is often referred to as the "master gland" because it controls the functions of many other endocrine glands. It is divided into two parts: the anterior pituitary (adenohypophysis) and the posterior pituitary (neurohypophysis). Some of the hormones produced by the pituitary gland include:

- **Growth Hormone (GH):** Stimulates growth, cell reproduction, and tissue repair.
- **Prolactin (PRL):** Stimulates milk production in mammary glands.
- **Thyroid-Stimulating Hormone (TSH):** Stimulates the thyroid gland to produce thyroid hormones.
- **Adrenocorticotropic Hormone (ACTH):** Stimulates the adrenal glands to produce cortisol and other steroid hormones.
- **Follicle-Stimulating Hormone (FSH):** Regulates the growth and development of ovarian follicles in females

and sperm production in males.

- **Luteinizing Hormone (LH):** Triggers ovulation and stimulates the production of testosterone in males.

3. Thyroid Gland: The thyroid gland is located in the neck and produces thyroid hormones that regulate metabolism, growth, and development. The main hormones produced by the thyroid gland are:

- **Thyroxine (T4) and Triiodothyronine (T3):** These hormones play a crucial role in regulating metabolic rate, energy production, and overall cellular activity.

4. Parathyroid Glands: The parathyroid glands are small glands located near the thyroid gland. They produce parathyroid hormone (PTH), which regulates calcium and phosphate levels in the blood and bone.

5. Adrenal Glands: The adrenal glands are located on top of each kidney and produce a variety of hormones that help regulate stress response, metabolism, and electrolyte balance. The adrenal cortex and adrenal medulla produce different types of hormones:

- **Adrenal Cortex Hormones:** These include cortisol (stress hormone), aldosterone (regulates sodium and potassium balance), and sex hormones (e.g., androgens, estrogens).
- **Adrenal Medulla Hormones:** The medulla produces adrenaline (epinephrine) and noradrenaline (norepinephrine), which are involved in the "fight or flight" stress response.

6. Pancreas: The pancreas serves both endocrine and exocrine functions. The endocrine portion of the pancreas produces hormones that regulate blood sugar levels:

- **Insulin:** Lowers blood glucose levels by promoting the uptake and storage of glucose in cells.
- **Glucagon:** Raises blood glucose levels by promoting the

release of glucose from glycogen stores in the liver.

7. Gonads (Ovaries and Testes): The gonads, which include the ovaries in females and the testes in males, produce sex hormones that regulate reproductive functions and secondary sexual characteristics:

- **Ovaries:** Produce estrogen and progesterone, which regulate the menstrual cycle and female reproductive system.
- **Testes:** Produce testosterone, which regulates male reproductive functions and secondary sexual characteristics.

These are just a few examples of the major endocrine glands and their associated hormones. The endocrine system's complex network of glands and hormones plays a crucial role in maintaining homeostasis and coordinating various physiological processes in the body.

Regulation of bodily functions through hormone secretion

Hormones are chemical messengers produced by endocrine glands that play a crucial role in regulating a wide range of bodily functions. These hormones are released into the bloodstream and travel to target cells or organs, where they exert their effects by binding to specific receptors. The regulation of bodily functions through hormone secretion involves complex feedback loops and interactions between different endocrine glands and systems. Here's an overview of how hormone secretion regulates various bodily functions:

1. Homeostasis: Hormones help maintain homeostasis, which is the body's ability to maintain stable internal conditions despite external changes. For example, the regulation of blood glucose levels by insulin and glucagon helps ensure that blood sugar remains within a certain range, which is essential for energy metabolism.

2. Metabolism: Hormones play a key role in regulating metabolism, which includes the processes of energy production, storage, and utilization. Thyroid hormones (T3 and T4) from the thyroid gland influence metabolic rate and energy expenditure, while insulin and glucagon regulate glucose metabolism.

3. Growth and Development: Hormones are essential for growth and development throughout life. Growth hormone (GH) from the pituitary gland promotes cell growth and division, while sex hormones such as estrogen and testosterone play a role in the growth and development of secondary sexual characteristics during puberty.

4. Reproduction: Reproductive hormones play a central role in regulating the reproductive system. Follicle-stimulating hormone (FSH) and luteinizing hormone (LH) from the pituitary gland regulate the menstrual cycle in females and sperm production in males. Estrogen and progesterone regulate the female reproductive cycle, while testosterone is crucial for male reproductive functions.

5. Stress Response: Hormones are involved in the body's response to stress. The adrenal glands release stress hormones, including cortisol, in response to various stressors. These hormones help the body cope with stress by increasing energy availability and modifying immune responses.

6. Blood Pressure and Fluid Balance: Hormones such as aldosterone from the adrenal glands and antidiuretic hormone (ADH) from the pituitary gland help regulate blood pressure and fluid balance by controlling sodium and water reabsorption in the kidneys.

7. Calcium Regulation: Parathyroid hormone (PTH) and calcitonin regulate calcium levels in the blood. PTH increases calcium levels by promoting its release from bone and increasing its absorption in the intestines. Calcitonin, produced by the thyroid gland, decreases blood calcium levels by promoting calcium deposition in bone.

8. Circadian Rhythms: The hormone melatonin, produced by the pineal gland, helps regulate sleep-wake cycles and circadian rhythms.

9. Immune Function: Some hormones, such as cortisol, have immunomodulatory effects and play a role in regulating immune responses to infections and inflammation.

10. Mood and Emotion: Neurotransmitters and hormones like serotonin, dopamine, and oxytocin influence mood, emotions, and social behavior.

These examples highlight the diverse ways in which hormone secretion regulates various bodily functions. The endocrine system's intricate communication network ensures that different systems work together to maintain optimal health and function.

Interplay between the endocrine and nervous systems

The endocrine and nervous systems are two major communication systems in the body that work together to regulate various physiological processes and maintain homeostasis. While they have distinct functions, there is a significant interplay between these systems, allowing them to coordinate responses to internal and external stimuli. Here's how the endocrine and nervous systems interact:

1. Communication and Coordination: Both systems are responsible for transmitting signals and information throughout the body to maintain balance and respond to changes. The nervous system uses electrical impulses to transmit signals rapidly, while the endocrine system uses chemical messengers (hormones) to transmit signals more slowly but with longer-lasting effects.

2. Hypothalamus-Pituitary Axis: The hypothalamus, a part of the brain, serves as a bridge between the endocrine and nervous systems. It controls the release of hormones from the pituitary gland by producing releasing and inhibiting hormones. These hormones regulate the secretion of various hormones from the anterior pituitary gland, which in turn controls the activity of other endocrine glands throughout the body.

3. Stress Response: The interplay between the endocrine and nervous systems is particularly evident in the body's response to stress. The hypothalamus activates the sympathetic nervous system ("fight or flight" response) and triggers the release of stress hormones such as adrenaline (epinephrine) and cortisol. Adrenaline is released from the adrenal medulla, while cortisol is released from the adrenal cortex. Together, these hormones prepare the body for a rapid response to stressors.

4. Feedback Loops: Both systems use feedback loops to regulate hormone secretion and maintain homeostasis. Negative feedback loops are common, where hormone levels influence the release of hormones from the pituitary gland or other endocrine glands. For example, the hypothalamus-pituitary-thyroid axis regulates thyroid hormone levels in response to changes in metabolic rate.

5. Hormonal Regulation of Neurotransmitters: Some hormones produced by the endocrine system can influence neurotransmitter activity in the nervous system. For instance, hormones like serotonin and dopamine, which affect mood and emotions, are influenced by endocrine processes.

6. Growth and Development: Both systems play roles in growth and development. Growth hormone (GH) from the pituitary gland promotes growth, while sex hormones produced by the endocrine system influence the development of secondary sexual characteristics during puberty.

7. Circadian Rhythms: The hypothalamus, particularly the suprachiasmatic nucleus, regulates the body's circadian rhythms and sleep-wake cycles. These rhythms are influenced by the interplay between the endocrine and nervous systems.

In summary, the endocrine and nervous systems are intricately interconnected and collaborate to regulate a wide range of physiological processes. Their interplay allows for rapid responses to immediate threats (nervous system) as well as more prolonged adjustments to maintain homeostasis (endocrine system). This integration ensures that the body can adapt to changing conditions and maintain optimal function.

Endocrine disorders and their impact on health

Endocrine disorders are conditions that result from dysfunction in the endocrine system, leading to imbalances in hormone production, secretion, or signaling. These disorders can have a significant impact on various physiological processes and overall health. Here are some common endocrine disorders and their potential impacts:

1. Diabetes Mellitus: Diabetes is a chronic metabolic disorder characterized by high blood sugar levels. It occurs when the body either doesn't produce enough insulin (Type 1 diabetes) or doesn't use insulin effectively (Type 2 diabetes). Uncontrolled diabetes can lead to:

- Cardiovascular Complications: Increased risk of heart disease, stroke, and hypertension.
- Nerve Damage (Neuropathy): Numbness, tingling, and pain in the extremities.
- Kidney Damage (Nephropathy): Kidney dysfunction and eventual kidney failure.
- Eye Complications (Retinopathy): Vision problems and even blindness.
- Foot Complications: Poor circulation and increased risk of infections, leading to foot ulcers and amputation.

2. Hypothyroidism: Hypothyroidism occurs when the thyroid gland doesn't produce enough thyroid hormones. It can lead to:

- Fatigue and Weakness: Decreased metabolism and energy levels.
- Weight Gain: Slower metabolism and fluid retention.

- Cold Sensitivity: Reduced heat production.
- Dry Skin and Hair: Decreased thyroid hormone affects skin and hair health.
- Mood Changes: Depression and cognitive impairment.

3. Hyperthyroidism: Hyperthyroidism results from an overactive thyroid gland producing excessive thyroid hormones. It can cause:

- Weight Loss: Increased metabolism leads to weight loss.
- Rapid Heartbeat: Elevated heart rate and palpitations.
- Heat Sensitivity: Increased heat production.
- Tremors and Nervousness: Nervous system stimulation.
- Eye Changes (Graves' Disease): Bulging eyes and eye discomfort.

4. Adrenal Disorders: Disorders of the adrenal glands can result in conditions such as Addison's disease (adrenal insufficiency) or Cushing's syndrome (excessive cortisol production). These can lead to:

- Addison's Disease: Fatigue, weakness, low blood pressure, and weight loss.
- Cushing's Syndrome: Weight gain, moon face, high blood pressure, and muscle weakness.

5. Polycystic Ovary Syndrome (PCOS): PCOS is a common hormonal disorder among women of reproductive age. It can lead to:

- Irregular Menstrual Cycles: Infrequent or absent periods.
- Ovulatory Dysfunction: Difficulty conceiving.
- Excess Androgen Production: Facial hair growth, acne, and hair thinning.

6. Growth Hormone Disorders: Disorders of growth hormone production can lead to:

- Dwarfism (Growth Hormone Deficiency): Stunted

growth and short stature.

- Gigantism (Excess Growth Hormone in Children): Overgrowth of bones and tissues.
- Acromegaly (Excess Growth Hormone in Adults): Enlarged facial features, hands, and feet.

These are just a few examples of endocrine disorders and their potential impacts on health. Endocrine disorders can affect multiple body systems and have wide-ranging consequences if left untreated. Timely diagnosis and appropriate management, often involving medication, lifestyle changes, or hormone replacement therapy, can help mitigate the impact of these disorders and improve quality of life. If you suspect you have an endocrine disorder, it's important to seek medical evaluation and guidance for proper diagnosis and treatment.

Male and female reproductive anatomy

Male and female reproductive anatomy refers to the structures and organs that are involved in the process of human reproduction. While there are differences between the reproductive systems of males and females, both systems work together to ensure the continuation of the species through the production and fertilization of gametes (sperm and eggs). Here's an overview of the male and female reproductive anatomy:

Male Reproductive Anatomy:

1. Testes: The testes are the primary male reproductive organs responsible for producing sperm and testosterone, the male sex hormone.

2. Scrotum: The scrotum is a sac of skin that holds the testes outside the body. It helps regulate the temperature of the testes, which is important for sperm production.

3. Epididymis: The epididymis is a coiled tube that lies on the surface of each testis. It functions to store and mature sperm before they are ejaculated.

4. Vas Deferens: The vas deferens is a muscular tube that carries mature sperm from the epididymis to the urethra.

5. Seminal Vesicles, Prostate Gland, and Bulbourethral Glands: These accessory glands secrete fluids that combine with sperm to form semen. Seminal vesicles contribute the majority of the fluid in semen, while the prostate gland and bulbourethral glands provide additional components.

6. Urethra: The urethra serves as a passage for both urine and semen. During ejaculation, semen is released from the reproductive system through the urethra.

7. Penis: The penis is the male organ responsible for transferring sperm into the female reproductive tract during sexual intercourse.

Female Reproductive Anatomy:

1. Ovaries: The ovaries are the female reproductive organs that produce eggs (ova) and hormones like estrogen and progesterone.

2. Fallopian Tubes: The fallopian tubes are narrow tubes that extend from the ovaries to the uterus. They provide a pathway for the egg to travel from the ovary to the uterus. Fertilization typically occurs in the fallopian tubes.

3. Uterus: The uterus, also known as the womb, is where a fertilized egg implants and develops into a fetus during pregnancy. It is lined with the endometrium, which thickens in preparation for pregnancy and sheds during menstruation if pregnancy doesn't occur.

4. Cervix: The cervix is the lower narrow part of the uterus that connects to the vagina. It serves as a barrier between the uterus and the vagina.

5. Vagina: The vagina is a muscular tube that connects the cervix to the external genitalia. It functions in sexual intercourse and serves as the birth canal during childbirth.

6. External Genitalia (Vulva): The vulva includes the external female genitalia, including the mons pubis, labia majora, labia minora, clitoris, and vaginal opening.

7. Mammary Glands: The mammary glands, located in the breasts, produce milk for breastfeeding after childbirth.

These structures and organs work together to facilitate the processes of gamete production, fertilization, pregnancy, and childbirth in humans.

Reproductive processes, fertilization, and pregnancy

Reproductive processes, fertilization, and pregnancy are essential aspects of human reproduction. These processes involve the successful interaction between male and female reproductive systems, leading to the creation of new life. Here's an overview of how these processes occur:

Reproductive Processes:

1. **Gametogenesis:** Gametogenesis is the process of producing gametes (sperm and eggs) through meiosis. In males, spermatogenesis occurs in the testes, leading to the formation of sperm cells. In females, oogenesis occurs in the ovaries, resulting in the development of mature egg cells (ova).
2. **Ovulation:** Ovulation is the release of a mature egg from an ovary. It typically occurs around the middle of the menstrual cycle. The released egg travels into the fallopian tube, where it may be fertilized by sperm.
3. **Fertilization:** Fertilization is the fusion of a sperm cell with an egg cell, resulting in the formation of a fertilized egg (zygote). Fertilization usually occurs in the fallopian tube. The sperm's genetic material combines with the egg's genetic material, forming a complete set of chromosomes.

Fertilization:

1. **Sperm Journey:** Sperm are ejaculated into the vagina

during sexual intercourse. They travel through the cervix, uterus, and into the fallopian tubes. Only a small number of sperm reach the fallopian tubes, where the egg is located.

2. **Fusion of Gametes:** When a sperm penetrates the egg's outer layer, the egg becomes impenetrable to other sperm. The sperm's nucleus fuses with the egg's nucleus, forming a zygote with a complete set of chromosomes.

3. **Zygote Formation:** The zygote undergoes several divisions, forming a blastocyst. The blastocyst travels down the fallopian tube and reaches the uterus, where it may implant into the uterine lining.

Pregnancy:

1. **Implantation:** The blastocyst attaches itself to the uterine lining, a process known as implantation. Once implanted, the blastocyst receives nourishment from the mother's blood vessels.

2. **Embryonic Development:** After implantation, the blastocyst develops into an embryo. This marks the beginning of pregnancy. The embryo undergoes various stages of development, including the formation of major organ systems.

3. **Fetal Development:** Around the eighth week of pregnancy, the embryo is referred to as a fetus. The fetus continues to grow and develop, with organs maturing and becoming functional.

4. **Pregnancy Trimesters:** Pregnancy is divided into three trimesters. Each trimester brings different developmental milestones for the fetus and various physical changes for the mother.

5. **Labor and Birth:** Towards the end of pregnancy, contractions begin, leading to labor. The cervix dilates, and the baby moves through the birth canal during

childbirth. After delivery, the placenta is expelled.

Pregnancy is a complex and remarkable process that involves the interplay of various physiological changes and hormonal interactions. Adequate prenatal care, proper nutrition, and regular medical check-ups are essential to ensure a healthy pregnancy and successful childbirth.

Hormonal regulation of the reproductive system

Hormonal regulation plays a crucial role in the functioning of the male and female reproductive systems. Hormones produced by various endocrine glands help coordinate and control processes such as gametogenesis (production of sperm and eggs), the menstrual cycle, ovulation, and pregnancy. Here's an overview of the hormonal regulation of the reproductive system in both males and females:

Hormonal Regulation in Males:

1. **Gonadotropin-Releasing Hormone (GnRH):** Produced by the hypothalamus, GnRH stimulates the anterior pituitary gland to release follicle-stimulating hormone (FSH) and luteinizing hormone (LH).
2. **Follicle-Stimulating Hormone (FSH):** In males, FSH stimulates the Sertoli cells in the testes to support sperm production (spermatogenesis). Sertoli cells also produce inhibin, which helps regulate FSH levels through negative feedback.
3. **Luteinizing Hormone (LH):** LH stimulates the Leydig cells in the testes to produce testosterone. Testosterone is crucial for the development and maturation of sperm, as well as the development of male secondary sexual characteristics.
4. **Testosterone:** Produced by the testes, testosterone is the primary male sex hormone. It plays a role in promoting sperm production, maintaining libido, regulating bone

density, and developing male physical traits.

Hormonal Regulation in Females:

1. **Gonadotropin-Releasing Hormone (GnRH):** As in males, GnRH from the hypothalamus stimulates the anterior pituitary gland to release FSH and LH.
2. **Follicle-Stimulating Hormone (FSH):** In females, FSH stimulates the growth and development of ovarian follicles, which contain maturing eggs (ova). The follicles also produce estrogen.
3. **Luteinizing Hormone (LH):** LH surge triggers ovulation, the release of a mature egg from the ovary. After ovulation, the ruptured follicle transforms into the corpus luteum, which produces progesterone.
4. **Estrogen:** Produced primarily by the ovaries, estrogen plays a central role in the menstrual cycle. It stimulates the thickening of the uterine lining (endometrium) in preparation for pregnancy and supports the development of secondary sexual characteristics.
5. **Progesterone:** Produced by the corpus luteum after ovulation, progesterone maintains the uterine lining, preparing it for embryo implantation and supporting early pregnancy. If pregnancy does not occur, progesterone levels drop, leading to menstruation.
6. **Human Chorionic Gonadotropin (hCG):** Produced by the developing placenta after embryo implantation, hCG maintains the corpus luteum during early pregnancy, ensuring the continued production of progesterone.

Hormonal Regulation During Pregnancy:

1. **Estrogen and Progesterone:** During pregnancy, the placenta becomes a significant source of estrogen and progesterone. These hormones help maintain the uterine environment, support fetal growth, and prepare the body for childbirth.

2. **Prolactin:** Produced by the pituitary gland, prolactin stimulates the development of mammary glands and milk production in preparation for breastfeeding.

Hormonal regulation in the reproductive system is a finely tuned process that involves feedback loops and interactions between various glands and hormones. These interactions ensure proper functioning of the reproductive processes in both males and females and are essential for fertility, pregnancy, and overall reproductive health.

Reproductive health and family planning

Reproductive health and family planning are important aspects of individual well-being, public health, and the overall quality of life. Reproductive health encompasses the physical, emotional, and social well-being related to all aspects of reproduction, while family planning involves making informed decisions about when to have children and how many children to have. Both concepts play a significant role in promoting healthy families, preventing unintended pregnancies, and ensuring the well-being of individuals and communities. Here's an overview:

Reproductive Health:

1. **Access to Comprehensive Healthcare:** Reproductive health includes access to quality healthcare services that address issues such as contraception, preconception care, prenatal care, childbirth, and postpartum care.

2. **Sexual Education and Awareness:** Providing accurate and comprehensive sexual education helps individuals make informed decisions about their sexual health, reducing the risk of sexually transmitted infections (STIs) and unintended pregnancies.

3. **Prevention of STIs:** Reproductive health emphasizes preventing the transmission of sexually transmitted infections through safe sexual practices, education, and awareness.

4. **Maternal and Child Health:** Ensuring safe pregnancies, childbirth, and postpartum care is vital for the health of both mothers and infants. This includes proper nutrition, prenatal care, and skilled attendance during

childbirth.

5. **Access to Contraception:** Reproductive health services provide access to a range of contraceptive methods to help individuals and couples plan when to have children, prevent unintended pregnancies, and space pregnancies.

6. **Safe Abortion Services:** Reproductive health includes access to safe and legal abortion services where permitted by law, along with counseling and support for those facing such decisions.

7. **Gender Equality and Empowerment:** Promoting gender equality and women's empowerment is essential for reproductive health. Empowered women are more likely to make informed choices about their reproductive health and family planning.

Family Planning:

1. **Birth Control Methods:** Family planning involves choosing and using various birth control methods to prevent or space pregnancies. These methods include hormonal methods (birth control pills, patches, injections), barrier methods (condoms, diaphragms), intrauterine devices (IUDs), and permanent methods (tubal ligation, vasectomy).

2. **Fertility Awareness:** Some couples use fertility awareness-based methods to track ovulation and plan or avoid pregnancy based on their fertile days.

3. **Emergency Contraception:** Emergency contraception methods can be used to prevent pregnancy after unprotected intercourse or contraceptive failure.

4. **Reproductive Rights:** Family planning is linked to reproductive rights, which include the right to decide the number and spacing of children, the right to access contraception, and the right to access safe and legal abortion services where permitted by law.

5. **Counseling and Support:** Family planning services often provide counseling and support to individuals and couples to help them make informed decisions about family size and spacing.

Promoting reproductive health and family planning contributes to healthier pregnancies, reduced maternal and infant mortality, improved economic opportunities for women, and the overall well-being of families and communities. Access to accurate information, comprehensive healthcare services, and the ability to make informed choices about reproductive options are essential for achieving these goals.

Structure and functions of the skin

The skin is the largest organ of the human body and serves as a protective barrier between the internal organs and the external environment. It plays a vital role in maintaining homeostasis, protecting against pathogens, regulating body temperature, and providing sensory information. The skin consists of multiple layers and structures, each with specific functions. Here's an overview of the structure and functions of the skin:

Structure of the Skin:

The skin is composed of three main layers:

1. **Epidermis:** The epidermis is the outermost layer of the skin and consists of several sublayers. It provides protection against UV radiation, pathogens, and chemicals. The outermost layer of the epidermis, called the stratum corneum, is composed of dead skin cells that provide a waterproof barrier.
2. **Dermis:** The dermis is the middle layer of the skin and contains various structures such as blood vessels, hair follicles, sweat glands, and nerve endings. It provides structural support and elasticity to the skin.
3. **Hypodermis (Subcutaneous Tissue):** The hypodermis is the deepest layer of the skin and is composed of adipose tissue (fat) that helps insulate the body and provides cushioning.

Functions of the Skin:

1. **Protection:** The skin acts as a physical barrier that protects the body from external factors such

as pathogens, harmful chemicals, UV radiation, and mechanical injuries.

2. **Regulation of Body Temperature:** The skin helps regulate body temperature through processes like sweating (cooling down the body) and vasoconstriction or vasodilation (regulating blood flow to control heat loss or retention).

3. **Sensation:** Nerve endings in the skin detect various sensations such as touch, pressure, pain, and temperature. These sensory receptors provide information about the external environment to the brain.

4. **Excretion:** Sweat glands in the skin secrete sweat, which helps eliminate waste products like urea and excess salts from the body.

5. **Immune Defense:** The skin's immune system defends against invading pathogens by preventing their entry and releasing antimicrobial substances.

6. **Synthesis of Vitamin D:** The skin plays a role in synthesizing vitamin D when exposed to UVB radiation from sunlight. Vitamin D is essential for calcium absorption and bone health.

7. **Blood Supply:** Blood vessels in the dermis provide nutrients and oxygen to the skin cells, helping maintain their health and function.

8. **Hair and Nails:** The skin contains hair follicles that produce hair and sebaceous glands that secrete oil to moisturize the skin and hair. Nails, composed of keratin, protect the tips of the fingers and toes.

9. **Absorption:** The skin can absorb certain substances, although to a limited extent. Transdermal patches and topical medications are examples of how substances can be absorbed through the skin.

The skin's structure and functions make it a complex and dynamic organ that plays a crucial role in maintaining the body's overall health and well-being.

Importance of the skin as a protective barrier

The skin serves as a critical protective barrier that plays a vital role in safeguarding the body against various external threats and maintaining overall health. Its role as a barrier is multifaceted and essential for the body's well-being. Here are some key reasons why the skin's function as a protective barrier is of paramount importance:

1. Defense Against Pathogens: The skin forms a physical barrier that prevents the entry of harmful microorganisms such as bacteria, viruses, and fungi. The outermost layer of the skin, the stratum corneum, consists of tightly packed dead skin cells that create an effective barrier against the penetration of pathogens and their toxins.

2. Prevention of Infections: By preventing pathogens from entering the body, the skin reduces the risk of infections. Even minor cuts and abrasions trigger the skin's repair mechanisms to heal the damaged area and prevent pathogens from entering the bloodstream.

3. Chemical and Environmental Protection: The skin provides protection against various harmful chemicals, pollutants, allergens, and irritants present in the environment. It acts as a shield, reducing the potential for these substances to penetrate and cause harm to internal organs and tissues.

4. UV Radiation Protection: The skin's pigment-producing cells, called melanocytes, produce melanin, which provides protection against the damaging effects of ultraviolet (UV) radiation from the sun. Melanin absorbs and scatters UV rays, preventing them from penetrating deeper layers of the skin and causing DNA damage.

5. Temperature Regulation: The skin plays a role in temperature regulation by preventing excessive heat loss from the body and preventing the entry of excessive heat from the environment. This is achieved through processes such as sweating and blood vessel dilation or constriction.

6. Sensory Functions: The skin contains sensory receptors that detect touch, pressure, pain, and temperature. These receptors provide essential information about the external environment, helping individuals avoid potentially harmful situations.

7. Prevention of Dehydration: The stratum corneum also acts as a barrier against water loss from the body, preventing dehydration. It helps retain moisture and prevents excessive evaporation of water from the skin's surface.

8. Physical Protection: The skin's physical structure and thickness provide a protective shield against mechanical injuries, such as cuts, scrapes, and impact. This is particularly evident in areas of the body exposed to potential trauma.

9. Immune Response: The skin's immune system responds to breaches in the barrier by releasing antimicrobial substances and immune cells to fight off potential infections.

In summary, the skin's function as a protective barrier is essential for maintaining the body's health and well-being. It guards against infections, physical injuries, environmental threats, and harmful substances, while also contributing to temperature regulation and sensory perception. Proper care and maintenance of the skin are crucial to ensure its optimal barrier function and overall health.

Role of sweat glands, hair, and nails in maintaining homeostasis

Sweat glands, hair, and nails play important roles in maintaining homeostasis, which is the body's ability to regulate internal conditions and balance in response to external changes. Each of these structures contributes to different aspects of homeostasis, including temperature regulation, protection, and sensory perception. Here's how sweat glands, hair, and nails contribute to maintaining homeostasis:

Sweat Glands:

1. **Temperature Regulation:** Sweat glands help regulate body temperature by producing sweat in response to elevated body temperature. Sweat contains water and electrolytes, and as it evaporates from the skin's surface, it carries away heat, cooling the body.
2. **Fluid and Electrolyte Balance:** Sweat glands also play a role in maintaining fluid and electrolyte balance. Sweating helps eliminate excess salts, toxins, and waste products from the body, contributing to overall homeostasis.

Hair:

1. **Insulation:** Hair on the skin's surface helps insulate the body, providing a layer of protection against temperature changes. In cold conditions, hair traps a layer of air close to the skin, creating an insulating barrier that reduces heat loss.

2. **Temperature Regulation:** While not as efficient as sweating, tiny muscles attached to hair follicles, called arrector pili muscles, can cause the hair to stand upright (goosebumps) in response to cold or fear. This action traps a layer of air, providing additional insulation.

Nails:

1. **Protection:** Nails protect the fingertips and toes from injuries and trauma. They help prevent damage to the sensitive skin underneath and aid in gripping and manipulating objects.
2. **Sensory Perception:** Nails contribute to sensory perception by enhancing touch sensitivity. Nails are richly innervated, and their presence enhances the tactile sensation when touching surfaces.

Overall, sweat glands, hair, and nails work together to help the body maintain its internal equilibrium and respond to changes in the external environment. These structures play roles in temperature regulation, fluid balance, protection against injuries, and sensory perception, all of which contribute to the body's homeostasis and overall well-being.

Skin disorders and care

Skin disorders encompass a wide range of conditions that affect the skin's appearance, texture, and overall health. These disorders can be caused by various factors, including genetics, environmental factors, infections, immune system dysfunction, and more. Proper skin care is essential for maintaining healthy skin and preventing or managing skin disorders. Here are some common skin disorders and tips for skin care:

Common Skin Disorders:

1. **Acne:** Acne is a skin condition characterized by the presence of pimples, blackheads, whiteheads, and cysts. It is often caused by excess sebum production, clogged pores, and bacterial growth.
2. **Eczema (Atopic Dermatitis):** Eczema is a chronic inflammatory skin condition that leads to red, itchy, and irritated skin. It often occurs in response to triggers like allergens or irritants.
3. **Psoriasis:** Psoriasis is an autoimmune disorder that causes skin cells to multiply rapidly, leading to thick, scaly patches of skin. It can be accompanied by itching and discomfort.
4. **Rosacea:** Rosacea is a chronic inflammatory condition that primarily affects the face, causing redness, visible blood vessels, and sometimes acne-like bumps.
5. **Contact Dermatitis:** Contact dermatitis occurs when the skin comes into contact with irritants or allergens, leading to redness, itching, and rash at the contact site.
6. **Skin Infections:** Various infections, such as fungal infections (e.g., ringworm), bacterial infections (e.g.,

impetigo), and viral infections (e.g., cold sores), can affect the skin.

7. **Hives (Urticaria):** Hives are raised, itchy welts on the skin that are usually triggered by allergic reactions.

Skin Care Tips:

1. **Keep Skin Clean:** Wash your skin regularly with mild cleansers to remove dirt, oil, and sweat. Avoid over-washing, as it can strip the skin of natural oils.
2. **Moisturize:** Apply a suitable moisturizer to keep the skin hydrated and prevent dryness. Choose a moisturizer appropriate for your skin type.
3. **Sun Protection:** Use sunscreen with at least SPF 30 to protect your skin from harmful UV rays, which can cause sunburn, premature aging, and skin cancer.
4. **Healthy Diet:** Eat a balanced diet rich in fruits, vegetables, whole grains, and lean proteins. Nutrients like vitamins A, C, and E contribute to healthy skin.
5. **Stay Hydrated:** Drink plenty of water to maintain skin hydration and overall health.
6. **Avoid Harsh Products:** Use gentle skin care products that are free from harsh chemicals, fragrances, and dyes, especially if you have sensitive skin.
7. **Manage Stress:** Chronic stress can exacerbate skin conditions. Practice stress management techniques like exercise, meditation, and deep breathing.
8. **Avoid Scratching:** If you have itchy skin, try to avoid scratching, as it can worsen irritation and lead to infections.
9. **Consult a Dermatologist:** If you have a skin disorder, consult a dermatologist for proper diagnosis and treatment. They can recommend appropriate medications and skin care routines.
10. **Hygiene Practices:** Practice good hygiene, such as using clean towels and avoiding sharing personal items to

prevent the spread of infections.

Remember that every individual's skin is unique, so it's essential to tailor your skin care routine to your specific needs and concerns. If you have a diagnosed skin disorder, following your dermatologist's recommendations is crucial for effective management and care.

Visual system and the anatomy of the eye

The visual system is a complex network of structures that allow us to perceive and interpret visual information from our environment. The primary organ of the visual system is the eye, which captures light and converts it into electrical signals that the brain interprets as visual images. Here's an overview of the anatomy of the eye and its components:

Anatomy of the Eye:

1. **Cornea:** The cornea is the transparent, dome-shaped outermost layer of the eye. It helps focus light onto the retina and plays a significant role in vision.
2. **Iris:** The iris is the colored part of the eye surrounding the pupil. It controls the amount of light entering the eye by adjusting the size of the pupil.
3. **Pupil:** The pupil is the black circular opening in the center of the iris. It regulates the amount of light that enters the eye.
4. **Lens:** The lens is a flexible, clear structure located behind the iris and the pupil. It adjusts its shape to focus light onto the retina.
5. **Retina:** The retina is a light-sensitive layer at the back of the eye. It contains specialized cells called photoreceptors (rods and cones) that capture light and convert it into electrical signals.
6. **Optic Nerve:** The optic nerve is a bundle of nerve fibers that carries the electrical signals generated by the photoreceptors in the retina to the brain for processing.
7. **Vitreous Humor:** The vitreous humor is a gel-like substance that fills the space between the lens and the

retina, providing structural support to the eye.

8. **Sclera:** The sclera is the white, tough outer layer of the eye that provides protection and structural integrity.

9. **Choroid:** The choroid is a vascular layer between the retina and the sclera. It supplies blood to the retina and helps nourish the eye's structures.

Visual Pathway:

1. **Light Entering the Eye:** Light enters the eye through the cornea and the pupil. The cornea and lens focus the light onto the retina.

2. **Photoreceptor Activation:** The photoreceptor cells in the retina (rods and cones) capture the light energy and convert it into electrical signals.

3. **Signal Transmission:** The electrical signals generated by the photoreceptors are transmitted to other retinal cells and then to the ganglion cells.

4. **Optic Nerve Transmission:** The ganglion cells' axons converge to form the optic nerve, which carries the electrical signals to the brain's visual processing centers.

5. **Brain Interpretation:** The brain processes the electrical signals and interprets them as visual images, allowing us to perceive and understand our surroundings.

The visual system is a remarkable example of the intricate interaction between the eye's structures and the brain. It enables us to experience the world through the sense of sight, capturing and processing visual information to create a rich and detailed perception of our environment.

Auditory system and the anatomy of the ear

The auditory system is responsible for our sense of hearing, allowing us to perceive and interpret sound stimuli from our environment. The primary organs of the auditory system are the ears, which capture sound waves and convert them into electrical signals that the brain interprets as auditory sensations. Here's an overview of the anatomy of the ear and its components:

Anatomy of the Ear:

The ear is divided into three main parts: the outer ear, the middle ear, and the inner ear.

1. Outer Ear:

- **Pinna (Auricle):** The visible external part of the ear that collects and directs sound waves into the ear canal.
- **Ear Canal (External Auditory Canal):** The tube-like structure that extends from the pinna to the eardrum. It carries sound waves to the eardrum.

2. Middle Ear:

- **Eardrum (Tympanic Membrane):** The eardrum is a thin membrane that separates the outer ear from the middle ear. It vibrates when sound waves strike it.
- **Ossicles:** The middle ear contains three small bones called ossicles: the malleus (hammer), incus (anvil), and stapes (stirrup). These bones transmit vibrations from the eardrum to the inner ear.
- **Eustachian Tube:** This tube connects the middle ear to the back of the throat. It helps equalize air pressure on

both sides of the eardrum and allows drainage of fluids.

3. Inner Ear:

- **Cochlea:** The cochlea is a spiral-shaped, fluid-filled structure that is responsible for converting sound vibrations into electrical signals that the brain can interpret.
- **Vestibular System:** The inner ear also contains the vestibular system, which helps maintain balance and spatial orientation.

Auditory Pathway:

1. **Sound Wave Reception:** Sound waves enter the ear through the pinna and travel down the ear canal to reach the eardrum.
2. **Vibration Transmission:** The eardrum vibrates in response to the sound waves. These vibrations are then transmitted to the ossicles in the middle ear.
3. **Amplification:** The ossicles amplify the vibrations and transmit them to the oval window, a membrane-covered opening that leads to the cochlea.
4. **Cochlear Activation:** The vibrations cause fluid in the cochlea to move, stimulating hair cells (auditory receptor cells) within the cochlea.
5. **Electrical Signals:** Stimulation of the hair cells generates electrical signals that travel through the auditory nerve to the brain's auditory processing centers.
6. **Auditory Interpretation:** The brain processes the electrical signals and interprets them as sound, allowing us to perceive and understand auditory sensations.

The auditory system is a complex and intricate sensory system that enables us to experience the richness of the sounds in our environment. The interaction between the outer, middle, and inner ear, along with the brain's processing of auditory signals, ensures that we can hear and comprehend a wide range of sounds.

Gustatory and olfactory senses

The gustatory and olfactory senses are two of the primary senses responsible for our perception of taste and smell, respectively. These senses work together to allow us to experience and differentiate various flavors and scents in our environment. Here's an overview of the gustatory (taste) and olfactory (smell) senses:

Gustatory (Taste) Sense:

The gustatory sense refers to our ability to perceive different tastes. Taste receptors are located on the taste buds, which are specialized structures found on the tongue, the roof of the mouth, and the throat. There are five primary taste sensations:

1. **Sweet:** Associated with sugars and sweet substances. It can indicate the presence of energy-rich foods.
2. **Sour:** Associated with acids and acidic substances. It can signal the presence of potentially harmful or unripe foods.
3. **Salty:** Associated with salts and sodium-containing substances. It helps regulate salt intake for proper bodily functions.
4. **Bitter:** Associated with bitter compounds. It can indicate the presence of potential toxins or harmful substances.
5. **Umami:** Associated with the savory taste of glutamate and certain amino acids. It is often found in protein-rich foods.

Taste perception involves taste receptors binding to specific molecules in food, triggering nerve impulses that are transmitted to the brain. The brain processes these impulses, allowing us to identify and distinguish different tastes.

Olfactory (Smell) Sense:

The olfactory sense involves our ability to perceive different scents or odors. Olfactory receptors are located in the olfactory epithelium, a specialized tissue lining the nasal cavity. When we inhale, airborne odor molecules come into contact with olfactory receptors, initiating nerve signals that travel to the brain for interpretation.

Key points about the olfactory sense include:

1. **Diversity of Scents:** Humans can distinguish thousands of different scents, and our olfactory system is highly sensitive to a wide range of molecules.
2. **Emotional and Memory Associations:** The olfactory sense is closely linked to emotions and memories. Certain smells can evoke strong emotional responses or trigger memories from the past.
3. **Direct Pathway to the Brain:** Unlike other senses that involve relay stations in the thalamus before reaching the cortex, olfactory signals have a direct pathway to the brain's olfactory bulb, contributing to quick perception of smells.

Interplay Between Taste and Smell:

The gustatory and olfactory senses often work together to create the perception of flavor. Much of what we consider taste is actually influenced by smell. For example, when we eat, aromas released from the food travel up the back of the throat and reach the olfactory receptors in the nasal cavity. The brain combines information from taste buds and olfactory receptors to create a more comprehensive perception of flavor.

Overall, the gustatory and olfactory senses play essential roles in our sensory experiences, contributing to our enjoyment of food, ability to detect potential dangers, and our ability to form emotional connections through the scents we encounter.

Role of sensory systems in perception and communication

The sensory systems play a crucial role in our perception of the world around us and in facilitating effective communication. These systems allow us to gather information from the environment, process it, and communicate our experiences to others. Here's how sensory systems contribute to perception and communication:

1. Perception:

- **Sensory Input:** The sensory systems, including vision, hearing, touch, taste, and smell, collect sensory input from the environment through specialized receptors. Each sensory system detects specific types of stimuli, such as light, sound waves, pressure, and chemicals.
- **Sensory Processing:** The brain receives and processes sensory information from various sensory organs. This processing involves interpreting, organizing, and integrating sensory signals to form a coherent perception of the environment.
- **Multisensory Integration:** The brain often combines information from multiple sensory modalities to create a comprehensive perception. For example, seeing a person's facial expression, hearing their voice tone, and feeling their touch can collectively convey their emotional state.

2. Communication:

- **Nonverbal Communication:** Sensory cues, such as facial expressions, body language, and gestures, are vital components of nonverbal communication. They convey emotions, intentions, and social cues that complement verbal communication.
- **Auditory Communication:** The auditory system allows us to communicate through spoken language, music, and other sounds. Our ability to hear and interpret speech sounds enables effective verbal communication.
- **Visual Communication:** Vision is central to visual communication through written language, symbols, images, and visual aids. Visual cues, such as facial expressions and body movements, also enhance the meaning of verbal communication.
- **Tactile Communication:** The sense of touch enables tactile communication through gestures, handshakes, hugs, and other forms of physical contact. Tactile cues convey emotions, comfort, and connection.
- **Olfactory and Gustatory Communication:** Smell and taste can communicate information about food, danger, or social interactions. Smell and taste cues can trigger memories and emotions that influence communication.

3. Emotion and Expression:

- **Emotional Perception:** Sensory experiences, such as seeing a loved one's smile or hearing a soothing voice, trigger emotional responses. These emotional cues are essential for understanding others' feelings and forming emotional connections.
- **Emotional Expression:** The sensory systems allow us to express our emotions through facial expressions, body language, tone of voice, and touch. These expressions help convey our emotional states to others.

4. Adaptation and Survival:

- **Sensory Alerts:** Sensory systems alert us to potential threats or changes in the environment. For example, hearing a sudden loud noise or feeling a sharp pain can signal danger.
- **Navigation:** Vision, proprioception (sense of body position), and balance contribute to spatial awareness and navigation, enabling us to move safely through our environment.

In summary, the sensory systems are essential for perceiving and interacting with the world, facilitating effective communication, forming emotional connections, and ensuring our adaptation and survival. Our ability to process sensory information and communicate experiences enriches our interactions with others and enhances our overall understanding of the world.

Integration of different body systems for overall functioning

The human body is a complex and interconnected system composed of various organs and systems that work together to maintain overall health and function. Integration among different body systems is essential for harmonious functioning and homeostasis, which is the body's ability to maintain a stable internal environment despite external changes. Here's how different body systems integrate for overall functioning:

1. Nervous and Endocrine Systems:

- **Communication and Coordination:** The nervous system, including the brain and spinal cord, and the endocrine system, which releases hormones, work together to regulate bodily functions. The nervous system provides rapid communication through electrical signals, while the endocrine system releases hormones for longer-lasting regulation.

2. Circulatory and Respiratory Systems:

- **Oxygen Transport:** The circulatory system (heart, blood vessels) and the respiratory system (lungs) collaborate to deliver oxygen-rich blood to tissues and remove carbon dioxide. Oxygen is absorbed from the lungs into the bloodstream and transported by the circulatory system to cells, where it's used for energy production.

3. Digestive and Circulatory Systems:

- **Nutrient Transport:** The digestive system breaks down food into nutrients that are absorbed into the bloodstream. The circulatory system then transports these nutrients to cells for energy, growth, and repair.

4. Muscular and Skeletal Systems:

- **Movement and Support:** Muscles attach to bones and work with the skeletal system to provide movement and stability. Muscles contract, generating force that allows bones to move.

5. Nervous and Muscular Systems:

- **Voluntary Movement:** The nervous system sends signals to muscles, enabling voluntary movement. Sensory feedback from muscles informs the nervous system about muscle tension and body position.

6. Respiratory and Muscular Systems:

- **Breathing:** Muscles, such as the diaphragm and intercostal muscles, work with the respiratory system to enable breathing. Contraction of these muscles expands the chest cavity, drawing air into the lungs.

7. Immune and Lymphatic Systems:

- **Immune Defense:** The immune system defends against infections and diseases. The lymphatic system supports the immune response by transporting immune cells and fluids to infection sites.

8. Nervous and Cardiovascular Systems:

- **Blood Pressure Regulation:** The nervous system helps regulate heart rate and blood vessel diameter, contributing to blood pressure control. Signals from the brain affect heart rate and vessel constriction or dilation.

9. Urinary and Circulatory Systems:

- **Waste Removal:** The urinary system filters waste products from the bloodstream and expels them as urine. The circulatory system carries waste-laden blood to the kidneys for filtration.

10. Reproductive and Endocrine Systems:

- **Hormonal Regulation:** The endocrine system regulates reproductive processes, including the menstrual cycle, puberty, and pregnancy. Hormones from the endocrine system influence reproductive functions.

Integration among these systems ensures that the body's physiological processes are balanced and coordinated. Disruptions in one system can have ripple effects on others, affecting overall health. The interdependence of body systems highlights the intricate web of interactions that underlie our well-being and functioning.

Case studies highlighting the interactions between systems

Here are two case studies that highlight the interactions between different body systems:

Case Study 1: Cardiovascular and Respiratory System Interaction

Scenario: John, a 50-year-old man, experiences shortness of breath and chest pain during exercise.

Interactions:

- **Cardiovascular System:** The heart pumps oxygenated blood to tissues and returns deoxygenated blood to the lungs for oxygenation. In this case, the heart's ability to pump enough oxygenated blood to meet the body's demands is compromised.
- **Respiratory System:** The respiratory system supplies oxygen to the bloodstream and removes carbon dioxide. Shortness of breath indicates inadequate oxygen exchange, affecting the oxygen supply to the bloodstream.
- **Integration:** The cardiovascular system's ability to deliver oxygen-rich blood depends on the respiratory system's efficiency in oxygen exchange. Impaired oxygen exchange can lead to reduced oxygen supply to tissues, causing symptoms like shortness of breath and chest pain.

Case Study 2: Muscular and Skeletal System Interaction

Scenario: Sarah, a 30-year-old woman, experiences back pain after lifting a heavy object.

Interactions:

- **Muscular System:** Muscles generate force and movement. Lifting heavy objects requires muscles to contract forcefully, generating the necessary strength to lift the object.
- **Skeletal System:** Bones provide support, protection, and leverage for muscles. Proper posture and lifting technique are crucial to preventing strain on bones and muscles.
- **Integration:** When lifting a heavy object, the muscular system contracts muscles to generate force, while the skeletal system provides structural support. If lifting technique is incorrect or excessive force is applied, it can strain muscles, ligaments, and bones, leading to back pain.

In both case studies, the interactions between systems are evident. The ability of one system to function effectively relies on the support and coordination of other systems. Disruptions or imbalances in one system can impact the overall functioning of the body and lead to symptoms or health issues. These case studies highlight the importance of understanding how different systems collaborate to maintain health and well-being.

The importance of holistic understanding in medical practice

Holistic understanding in medical practice refers to the approach of considering the entire patient and their individual needs, rather than focusing solely on the disease or symptoms. It emphasizes treating the patient as a whole, taking into account physical, emotional, social, and psychological aspects of their health. Here's why holistic understanding is crucial in medical practice:

1. Comprehensive Patient Care: A holistic approach considers all aspects of a patient's well-being, including physical, mental, emotional, and social factors. This helps address the root causes of health issues and provides more comprehensive and effective care.

2. Individualized Treatment: Every patient is unique, with different genetics, lifestyles, and circumstances. A holistic approach tailors treatments to the individual, taking into account their specific needs and preferences.

3. Addressing Underlying Causes: Holistic understanding goes beyond treating symptoms. It seeks to identify and address the underlying causes of health issues, which can lead to more effective and lasting outcomes.

4. Prevention and Wellness: By considering all aspects of a patient's health, holistic medicine focuses on preventive measures and promoting overall wellness, which can reduce the risk of future health problems.

5. Improved Patient-Doctor Relationship: Holistic understanding fosters better communication and trust between patients and healthcare providers. Patients feel heard and understood, leading to better adherence to treatment plans.

6. Mental and Emotional Well-Being: Physical health is closely linked to mental and emotional well-being. Holistic care recognizes the impact of psychological factors on physical health and vice versa.

7. Chronic Disease Management: For chronic conditions, holistic care can help patients manage symptoms, improve their quality of life, and reduce the need for frequent medical interventions.

8. Long-Term Benefits: Holistic care focuses on long-term health rather than short-term fixes. This approach can lead to improved overall health and greater patient satisfaction.

9. Integrative Approaches: Holistic understanding encourages integration of complementary therapies, lifestyle modifications, and conventional treatments to achieve optimal results.

10. Patient Empowerment: Patients actively participate in their own care when a holistic approach is taken. They become more empowered to make positive changes in their lifestyles.

11. Cultural Sensitivity: Holistic understanding respects cultural beliefs and practices that can influence a patient's health and well-being.

12. Treating the Whole Person: Ultimately, holistic medicine recognizes that a person is not just a collection of organs and systems but a complex interplay of physical, emotional, social, and spiritual elements.

Incorporating holistic understanding in medical practice leads to more patient-centered care that considers the person as a whole, leading to better outcomes, improved patient satisfaction, and a more comprehensive approach to health and healing.

Reflection on the journey through human anatomy

The journey through human anatomy is a remarkable exploration that unveils the intricate complexity and interconnectedness of the human body. It's a journey that takes us from the microscopic realm of cells to the macroscopic structures that shape our form and function. Along this journey, we discover the awe-inspiring mechanisms that allow us to see, hear, move, breathe, digest, and experience the world in all its richness.

The study of human anatomy reveals the elegance of design, the efficiency of function, and the incredible adaptability that has enabled humans to thrive in diverse environments. From the framework of the skeletal system to the orchestration of the nervous system, every component has a role to play in maintaining equilibrium and facilitating survival.

As we delve deeper into the intricacies of anatomical structures, we come to realize that the body is not just a collection of parts, but a harmonious symphony of systems working together. The interplay between these systems highlights the importance of integration for overall health and well-being. The cardiovascular system pumps life-giving blood, the respiratory system sustains us with each breath, and the digestive system nourishes and fuels our every action.

Throughout this journey, we encounter the delicate balance between form and function. We appreciate the interdependence of organs, the complex networks of nerves, the elegance of muscular contractions, and the intricate dance of hormones. We marvel at the body's ability to heal, adapt, and respond to internal and external changes.

But the journey through human anatomy is not only a scientific exploration; it's also a humbling experience that evokes wonder and reverence. It's a reminder of our interconnectedness with the natural world and the intricate biological processes that allow us to exist.

As we conclude this journey, we are left with a profound appreciation for the intricacies of human anatomy and the role it plays in shaping our experiences as sentient beings. It's a journey that invites us to marvel at the beauty of life's design and the remarkable complexity that defines our existence.

The ongoing relevance of anatomical knowledge

Anatomical knowledge remains highly relevant and essential in various fields, ranging from medicine and healthcare to biology, sports science, and even art. Here's why anatomical knowledge continues to hold ongoing relevance:

1. Medical and Healthcare Fields:

- **Diagnosis and Treatment:** An understanding of anatomy is crucial for accurate diagnosis and effective treatment of medical conditions. Physicians, surgeons, and other healthcare professionals use anatomical knowledge to interpret symptoms, plan surgeries, and administer treatments.
- **Patient Care:** Healthcare providers use anatomical knowledge to provide holistic care and communicate effectively with patients about their conditions and treatment options.

2. Medical Education:

- **Training:** Medical students and professionals need a strong foundation in anatomy to develop clinical skills and perform procedures safely and accurately.
- **Advancements:** As medical technology and techniques advance, anatomical knowledge remains fundamental to integrating new methods into practice.

3. Surgical Procedures:

- **Precision:** Surgeons rely on anatomical knowledge to perform surgeries with precision and minimize risks to surrounding structures.
- **Innovation:** Advancements in surgical techniques, including minimally invasive procedures, require an in-depth understanding of anatomical landmarks and structures.

4. Research and Biomedical Sciences:

- **Disease Mechanisms:** Anatomical knowledge helps researchers understand how diseases affect different organs and tissues, contributing to the development of treatments and therapies.
- **Drug Development:** Understanding anatomical structures at a cellular level informs drug design and delivery methods.

5. Physical Therapy and Rehabilitation:

- **Recovery:** Anatomical knowledge guides physical therapists in designing customized rehabilitation programs that target specific muscles, joints, and structures.
- **Prevention:** Understanding biomechanics and anatomical limitations helps prevent injuries and optimize recovery.

6. Sports Science and Athletics:

- **Performance Enhancement:** Athletes and coaches benefit from understanding the anatomy of muscles and joints, aiding in injury prevention, skill development, and performance enhancement.

7. Art and Design:

- **Realism:** Artists, animators, and designers use anatomical knowledge to create accurate depictions of the human form, improving the realism and authenticity of their work.

8. Ergonomics and Product Design:

- **Product Development:** Anatomical knowledge informs the design of ergonomic furniture, tools, and equipment that are comfortable and safe for human use.

9. Forensic Science:

- **Identification:** Anatomical knowledge is vital in forensic investigations to identify remains, understand trauma, and reconstruct events.

10. Public Health and Education: - **Health Literacy:** Anatomical education enhances health literacy, empowering individuals to make informed decisions about their well-being. - **Preventive Measures:** Understanding anatomy helps promote healthy lifestyles and encourages preventive measures.

In essence, anatomical knowledge serves as the foundation for various disciplines, promoting advancements in healthcare, research, education, and creativity. As our understanding of human anatomy continues to deepen, its relevance and applications across diverse fields will remain indispensable.

Encouragement for further exploration and appreciation of the human body

Embarking on a journey of further exploration and appreciation of the human body is an endeavor that promises to reveal wonders beyond imagination. The intricate design, remarkable complexity, and harmonious interplay of systems within our bodies are an invitation to delve deeper into the mysteries of life itself. Here's an encouragement for your journey:

Unveil the Wonders: As you explore the human body, you'll uncover the wonders of cellular intricacies, the symphony of organs, the choreography of muscles, and the orchestra of nerves. Each discovery will ignite a sense of awe and amazement, inspiring you to seek further.

Inspiration for Learning: The human body serves as a living textbook, offering lessons in biology, physiology, and anatomy. Every organ, tissue, and cell is a page waiting to be read, providing insight into the mechanisms that sustain our existence.

Empowerment through Knowledge: As you understand the workings of your body, you gain the power to make informed choices about your health. Knowledge about nutrition, exercise, and self-care empowers you to nurture your well-being.

Bridge between Science and Art: Exploring the human body bridges the gap between science and art. It invites you to appreciate the beauty of biological design, whether you're a medical professional, artist, or simply someone curious about life's intricacies.

Curiosity Unleashed: Every corner of the human body invites your curiosity. Whether you're studying the brain's neural pathways, the heart's pumping action, or the elegance of joint movements, there's always something captivating to discover.

Deepen Self-Understanding: By understanding your body, you gain insight into your own unique strengths, vulnerabilities, and potential. You'll appreciate the body's resilience and its role in expressing emotions and experiences.

Fuel for Innovation: A journey into the human body fuels innovation. Medical breakthroughs, technological advancements, and therapies arise from understanding the body's inner workings.

Holistic Connection: As you study the body's systems, you'll witness how they interact to create a harmonious whole. This understanding mirrors the interconnectedness of life on a broader scale.

Lifetime of Exploration: The study of the human body is a lifelong endeavor. Even experts continue to uncover new insights, ensuring that there's always more to learn and explore.

Gift of Curiosity: The human body is a gift waiting to be unwrapped with curiosity. Embrace this gift, and let your exploration be a tribute to the wonder of existence.

So, embrace the invitation to explore and appreciate the human body. Whether you're a student, a healthcare professional, an artist, or simply a curious soul, this journey promises a world of revelations, insights, and a profound appreciation for the marvels that make us who we are.

Glossary of anatomical terms

Sure, here's a glossary of some common anatomical terms:

A

- **Abduction:** Movement of a limb away from the midline of the body.
- **Adduction:** Movement of a limb toward the midline of the body.
- **Anterior:** Toward the front or belly side of the body.
- **Articulation:** The point of connection between two bones, also known as a joint.

B

- **Bilateral:** Pertaining to both sides of the body.
- **Brachial:** Relating to the arm, specifically the upper arm.

C

- **Cartilage:** Tough, flexible connective tissue that cushions joints and supports body structures.
- **Cervical:** Relating to the neck or the neck region of the spine.
- **Cranial:** Pertaining to the head or skull.
- **Cranium:** The skull.

D

- **Dorsal:** Toward the back of the body.

E

- **Extension:** Straightening or increasing the angle

between body parts.
- **Flexion:** Bending or decreasing the angle between body parts.

F

- **Femoral:** Relating to the thigh or femur bone.
- **Fibula:** The smaller of the two lower leg bones.
- **Frontal Plane:** Divides the body into front and back portions.

H

- **Humerus:** The bone of the upper arm.
- **Hyoid Bone:** A U-shaped bone in the neck that supports the tongue.

I

- **Inferior:** Below or lower in position.
- **Intervertebral Disc:** A cushion-like structure between the vertebrae of the spine.

L

- **Lateral:** Away from the midline of the body.
- **Ligament:** Tough fibrous tissue that connects bones to other bones.
- **Lumbar:** Relating to the lower back or the lumbar region of the spine.

M

- **Medial:** Toward the midline of the body.
- **Metatarsals:** Bones in the foot between the ankle and the toes.
- **Muscle:** Tissue that contracts to produce movement.

N

- **Nasal:** Relating to the nose.

- **Neck:** The portion of the body connecting the head to the torso.
- **Nervous System:** The body's control system, including the brain, spinal cord, and nerves.

O

- **Occipital:** Relating to the back of the head or skull.
- **Organ:** A structure made up of tissues working together to perform specific functions.

P

- **Palmar:** Relating to the palm of the hand.
- **Patella:** The kneecap.
- **Pelvis:** The basin-shaped structure at the base of the spine.
- **Phalanges:** The bones of the fingers and toes.
- **Plantar:** Relating to the sole of the foot.
- **Posterior:** Toward the back or spine side of the body.
- **Proximal:** Nearer to the point of attachment or origin.
- **Pubic:** Relating to the pubis bone.

R

- **Radius:** The shorter of the two bones of the forearm.
- **Rectus:** Straight muscle fibers that run parallel to the body's midline.
- **Rotator Cuff:** A group of muscles and their tendons that stabilize the shoulder joint.

S

- **Sagittal Plane:** Divides the body into left and right portions.
- **Scapula:** The shoulder blade.
- **Skeletal System:** The body's framework of bones and associated tissues.
- **Superior:** Above or higher in position.

- **Synovial Joint:** A joint that allows movement, enclosed by a synovial capsule.

T

- **Tarsals:** The bones of the ankle.
- **Thoracic:** Relating to the chest or the thoracic region of the spine.
- **Tibia:** The larger of the two lower leg bones.
- **Tissue:** A group of cells that perform a specific function together.

U

- **Ulna:** The longer of the two bones of the forearm.

V

- **Vertebrae:** The individual bones of the spine.
- **Viscera:** Organs in the body's cavities, especially the abdominal cavity.

This glossary provides a snapshot of the vast anatomical terminology that describes the human body's structures and functions. It's a starting point for those interested in delving into the intricate language that conveys the complexity of our bodies.

Resources for additional reading and learning

Here are some reputable resources for additional reading and learning about human anatomy:

1. Books:

- **"Gray's Anatomy for Students" by Richard Drake, A. Wayne Vogl, and Adam W. M. Mitchell:** A widely respected textbook that covers human anatomy in detail, with clear illustrations and explanations.
- **"Netter's Atlas of Human Anatomy" by Frank H. Netter:** An atlas featuring beautifully detailed and labeled illustrations of human anatomy, helping you visualize structures.
- **"Human Anatomy & Physiology" by Elaine N. Marieb and Katja N. Hoehn:** A comprehensive textbook covering both anatomy and physiology, suitable for students and those interested in a detailed understanding.
- **"Anatomy: A Photographic Atlas" by Johannes W. Rohen, Chihiro Yokochi, and Elke Lütjen-Drecoll:** This atlas features photographs of real human dissections, offering a unique perspective on anatomy.

2. Online Resources:

- **Visible Body:** An interactive 3D anatomy visualization tool that allows you to explore the human body's structures and systems in detail.
- **InnerBody:** A website that provides interactive, educational tools to learn about the human body's

various systems and structures.

- **AnatomyZone:** Offers free video tutorials that break down complex anatomical concepts, making learning more accessible.
- **Anatomy.tv:** Offers 3D anatomical models, interactive quizzes, and educational resources for a deeper understanding of human anatomy.

3. Online Courses:

- **Coursera:** Offers various online courses on human anatomy, physiology, and related topics from universities and institutions around the world.
- **Khan Academy:** Provides free educational videos and articles on anatomy and physiology, suitable for learners of all ages.
- **edX:** Offers courses on anatomy, physiology, and related subjects from reputable universities and institutions.

4. Educational Institutions:

- **Local Universities and Colleges:** Many educational institutions offer anatomy courses as part of their curriculum. Check with local universities or community colleges for opportunities to enroll in classes.
- **Medical Schools:** Some medical schools offer public lectures, workshops, and seminars on anatomy topics. Keep an eye out for such events in your area.

5. Medical and Anatomy Museums:

- **Visit medical and anatomy museums:** Many cities have medical or science museums that feature exhibits on human anatomy. These exhibits often provide hands-on learning experiences.

When exploring these resources, remember that human anatomy is a vast subject, and different resources may cater to different levels of understanding. Whether you're a student, healthcare professional, artist, or simply curious, these resources can help deepen your knowledge and appreciation of the intricacies of the human body.